The Way It Was

by
Glenn E. Taylor

The Way It Was

Published by Carpenter's Son Publishing, Franklin, Tennessee

Published in association with Larry Carpenter of Christian Book Services, LLC
www.christianbookservices.com

Cover and Interior Layout Design: Suzanne Lawing

Printed in the United States of America

978-0-9839876-6-6

ACKNOWLEDGEMENTS

First, I give credit to my wife, Sarah Jane, who has been my anchor and most ardent supporter for more than 61 years. I also offer special praise to my daughter, Kathie Tharpe, who has been my advisor and "Business Manager" throughout the publication of the book. My other two children, Mitchell Taylor and Melinda Woller, have been advisors and "sounding boards" throughout the entire project.

Ned Handy, author of "The Flame Keeper," and Emo Maiden, both of the 466th Bomb Group era, got me to thinking that this story was worth publishing, if for no other reason than as a historical document. Gil Cohen, the artist that painted the beautiful rendition of "The Crewman," has honored me by giving permission to use his artwork on the cover of my book. I also would like to thank my publisher, Larry Carpenter, for his advice and guidance in this complicated process. To all these people, I am indebted.

Glenn E. Taylor

INTRODUCTION

Following my retirement from Norfolk Southern Railway on September 1, 1987, I was not prepared for the inactivity of retired life and started looking for something to do to occupy my time. I had always wanted to have a woodworking shop, but that took time and money. So I finally decided to buy a Smith Corona typewriter and start writing my life's story. But where to start? I had many experiences from which to draw: life on the farm, my military experience, involvement in several Chambers of Commerce in four locations in two states over fourteen years, or, finally, my twenty-three years with the railroad – Southern and Norfolk Southern.

Having decided on the military years, I engaged my new typewriter and delved into a world I hadn't thought much about for more than 40 years. What appealed to me the most, and I hoped to my family, was the record I could re-live on paper, telling them the harrowing stories of flying at twenty thousand feet at sub-zero temperatures, over the hostile skies of occupied Europe, during World War II.

When I finished that part of my life's story, I realized that I had to finish my total story by also writing the other events of my life: farm boy, businessman and railroader. I have not included the latter two stories in this account for fear it would detract from the history of a young boy growing up on an Antebellum plantation in Piedmont North Carolina, being caught up in the war frenzy and serving his country in the 1943-45 period. I will have more to say about my later years – college and business life – in the conclusion of this story. I hope you enjoy reading this historic document as much as I have enjoyed living it.

I never intended to publish this as a book, but the encouragement of family and friends has given me the inspiration to go for it! I offer it as a true-life record of a boy/man and the complexities he encountered.

I place one caveat on this work: it was written forty-five years after I experienced it, and the mind has a way of playing tricks over that length of time.

Much credit must be given to at least four people who have helped in its completion, and I would be remiss in not sharing credit with them. First, my wife, Sarah Jane Taylor, encouraged me to put my experiences on paper and has read and proofed the results several times. My daughter, Kathie Tharpe, typed one segment of the story and had it printed. My former secretary, Susan Bickford, was kind enough to type the original copy of the Fairntosh portion of my story. Lastly, my granddaughter, Virginia Tharpe Cochran, is due special thanks for retyping the entire epistle. To each of these, I owe a great deal and could never have brought my work to the point of completion without their assistance.

Although she does not know it, my cousin, Peggy (Taylor) Winstead, is probably more responsible for my getting the original manuscript out, dusting it off, and bringing it back to life than anyone else. After reading her book, *Peggy's Story*, I again became interested in my own – after over fifteen years.

Through the cooperation, help, and encouragement of these people, this story has been written and rewritten, proofed and re-proofed, read and reread several times. For what it is worth, the story that follows is mine.

Glenn E. Taylor
2011

SECTION I: THE EARLY YEARS
MY CHILDHOOD ADVENTURES

Rain poured torrents as we approached Stagville, our home for the next few months. The date was August 1931, and we had been on the road all day transporting the Taylor family "goods" from Dobson, North Carolina, to Fairntosh Plantation, a distance of some 85 miles. As we turned off the main road down a muddy, water-holed trail, amongst overhanging mock orange and tall elm trees, my heart sank. Lightning flashed and thunder rolled as the torrential rains fell on the nondescript road. Finally, after perhaps a half-mile, we approached the house at Stagville, a colonial mansion that was built around the turn of the 18th Century. In modern terms, it was small, but it was stately—perhaps eight rooms—some of them cubby-hole small and the others amazingly large. As the rain fell and the lighting flashed, I discerned a ghost-like home that would do credit to Wuthering Heights.

I rode in the truck that was carrying Mother and Dad's life-long accumulation of furniture. We had nothing remarkable because this was during the heart of the depression. We were fortunate that Dad had found a job after being "relieved" of his position as county agent of Surry County—due to political pressure. He had been an excellent county agent, but the opposing party had been elected, and he was out!

Stagville was one of several parts of Fairntosh Farms, which Dad had been hired to manage. Some 6,000 acres in all, the plantation embraced the dying of an old dynasty and the rebirth of a new. It was run down, decayed, and farmed by tenants who did not really understand modern farming principles. Dad was the answer to the problem. As a graduate of North Carolina State and a teacher of Vocational-Agriculture for several years prior to becoming county agent, he was well equipped to face the task of rejuvenating such an operation. It was a tough assignment, but he was eager for the challenge.

Fairntosh consisted of several sections in its vast acreage: Stagville, Fairntosh, Horton Grove, Harris, Eno Quarters and Boo Hoo, plus Red Mill and the "low grounds" on the river. Originally, the plantation had embraced Snow Hill and Brick House – 36,000 acres in all. However, much of it was lost to reconstruction taxes following the Civil War. Actually, Fairntosh was a dream come true for any little boy looking for an adventure and a back yard playground of thousands of acres. It was more than I could imagine, especially after Dobson, where we had a yard of a couple of acres.

Our New Life

It did not take us long to settle in and become one with the plantation. The first fall, V.C. (my older brother), and I acclimated ourselves to the Stagville area. Meanwhile, Dad hired carpenters to renovate the manager's house for livability and restore the big barn and out buildings for their proper use.

We little fellows explored the hills and gullies, the old Bennehan Cemetery, the pastures and wild grapevines that hung in profusion down the hill below the house. Stagville had a cistern, long unused and a barn that was constructed so the hay wagons could be driven to the loft area for unloading while the livestock was stabled underneath the slope of the hill. The lower story was of rock masonry while the upper level (loft) was of wood siding, styled after many Virginia barns.

Dad's first purchase was a pair of iron gray Percherons, named Pearl and Nell, followed by the purchase of a saddle horse named Nancy. Years later, she would become my saddle horse and my first love. Dad also bought an F-20 Farmall tractor with steel-cleated wheels to break ground and begin fall planting. I vividly recall that first fall and how he asked me to accompany him to Fairntosh on a horse-drawn mowing machine to deliver the livestock and machine to the barn. We took a direct route through primeval forests of hardwood that had never been cut. The small road, almost a path hewn out of the forest, was overlapped by massive oak and hickory causing the evening darkness to settle even quicker than ordinary. I sat on the tongue of the mower between Dad's legs, and only through his presence, did I feel a sense of security. I had never experienced a body of timber so awesome, but then in 1931, I was only seven years old.

As I grew and became accustomed to those woods, they became my play ground and later, my favorite hunting forest.

Our New Home

Perhaps three months after moving to Stagville, our home at Fairntosh was renovated and we moved. The manager's home was a typical farmhouse with four rooms in front, two rooms down and two up, divided by a large hall and stairs. The back of the house was in a "T" shape that accommodated the dining room and the kitchen. Each room had a fireplace. Across the front was a full width one-story porch, and down the right side was a long screened porch with a bath built onto the back. Behind the kitchen was a smaller porch.

One of the first priorities was to develop a running water system from the well in the side yard. The well was cleaned, and a new pump house was built over it. Inside, we had a one hundred gallon tank with a half-horse power gasoline engine that powered the well pump. Pipes ran from the well to the bathroom and kitchen supplying our water. The water was heated by a wood range in the kitchen with a water jacket attachment on the side through which the water flowed. Not having electricity, we made the most of the innovations that were available at the time.

When the water system was completed, one of the workers was determined to have the first drink out of the new pipes. He probably had never seen a running water spigot in his life and rushed around behind the well house where Dad had installed an extra spigot for the outdoor laundry, turned on the spigot and drank of its contents. The devil made him do it, but Dad said, "Wert, did you drink that water?" The reply was a proud "Yessa." Dad picked up an old can of KRO rat poison that had probably been at the well house for years and showed it to Wert, who could not read, but understood the skull and crossbones label on the can. Convinced that he had been poisoned, he asked if there was any cure, to which Dad replied that perhaps a dose of castor oil would "wash out" his poison. He hurried to the back door and asked Mother if she had any castor oil, which she did, and he took a ten-cent bottle (a lot in those days) and drained it. He was excused for the balance of the day and hurried home to bid his family farewell. About halfway home, the castor oil began to work, and Wert was sure that he was going to die without ever seeing his sweet wife Mandy again.

Keneycut Club vs. Saltycuts

I well remember so many things that happened that first fall. Pearl died of "lock-bowels" and had to be dragged off and buried. Due to her enormous size, the hired hands spent days digging a grave big enough to accommodate her. This was my third year in school and being the "new kid on the block," I was constantly harangued, beaten, and mocked. The three years I went to the Wiley P. Mangum School in the nearby community of Bahama were not happy ones. One little redhead named Grover Ellis was a feisty little cuss and beat up on me every day as a workout before he took on the big boys. He was killed at Guadalcanal in World War II. I am sure he left a mound of Japanese in front of him before he met his fate.

Speaking of Bahama, my brother, V.C., was having the same problem of being accepted in his class, as were several others. We formed the "Keneycut Club" to protect ourselves from the bully element, which might have worked except the bullies formed the "Saltycuts." We had regular gang warfare that ranged between the schoolhouse and the N&W Railroad tracks, and from Bahama proper to as far as we could run and get back to school before bus time. Forget class! At one juncture, the Saltycuts captured Bruce Pope (better known as "Poke 'n Beans") and tied him to the railroad tracks. Next, they captured "Winkey" Lloyd and swung him by the ankles, upside-down, from the top of a sapling. They were tough, but then, we were tough too! V.C. and I cut a dogwood tree, perhaps three or four inches at the stump, that had a distinct "Y," carried it to school and planted it behind our battle lines. With the help of a truck inner tube, we could lob softball-sized rocks into the enemy lines. Fortunately, no one was ever hit. The principal could not handle the bunch of ruffians, but armistice was declared when Dad got wind of the fracas and took the "bull by the horns." Both the Keneycuts and Saltycuts were out of business in short order. I fell in and out of love two or three times in the third and fourth grades and won a couple of fights, while losing fifty.

FAIRNTOSH

Meanwhile, back at Fairntosh, I did what most little boys do – I played a lot and was imaginative in my play. Behind the "Big House," the antebellum mansion that sat in the center of a forty-acre lawn, was a large wisteria vine that at one time had been the center of a formal garden. In the early thirties, Tarzan was the idol of most little boys, and we were no exception. We would swing through the wisteria vines and yodel to our hearts' content. Winkey Lloyd was our closest playmate, and he would come over occasionally to spend the afternoon or night. We invited him to swing and yodel in the truest Tarzan fashion; however, Winkey was not too coordinated. When he sprang from the first vine to the second and missed, his Tarzan yell became, "Ie, ei, oh, uh, oh!" Mother got tired of hearing the yelling all over the lawn and threatened never to allow us to see another movie if she heard the Tarzan yell one more time!

A Special Place

Fairntosh encompassed six thousand acres broken into several farm sections by colorful names. The farm was bordered on three sides by the Little River that flowed into the Eno on one side and the Flat River on the other. These met the Eno to form the Neuse River. Only the north was unbordered by a river. There were several huge tracts of virgin timber throughout the farm, little of which was cut while we lived there. Vast forests of original growth pine, seventy-five or more feet high with trunks four feet in diameter, stood majestically in two tracts of perhaps five hundred to a thousand acres each. Moreover, the swampland along the Eno contained hundreds of acres of hardwood and gum - some of the most gigantic trees I have ever seen in the southeast. In Boo-Hoo, another thousand acres of hardwood stood untouched until sometime after we moved there. Some of the timber was cut for market; however, much of it was used for construction of homes, barns, and outbuildings for the use of the farm.

Almost every year we lived there, the rivers would flood, sometime to the extent that we were marooned for a day or so, except by way of the community of Bahama. When these floods occurred during the summer months, they could cause extensive crop damage – loss of grain, corn, hay, etc. – but the benefits of the rich bottomland far outweighed the handicaps. I have harvested fields of oats that yielded 55 bushels to the acre and wheat that produced upward of 50 bushels. The corn crop was likewise bountiful in a good year as was the hay crop. The low-grounds in the spring of the year, before the grain harvest and after the corn was knee high, were a beautiful sight. Hundreds of acres of golden grain and black-green corn waving in the gentle breeze were enough to confirm the words of "America the Beautiful."

Rocky Ridge

Rocky Ridge was a phenomenon of nature at Fairntosh. It was a small ridge of rock perhaps three or four hundred yards wide and two miles long that was interrupted in only one location. Its height was no more than a couple of hundred feet, and it contained a profusion of oak and hickory trees—the fruit that squirrels love to eat. Moreover, the boulders were spaced just so a person could jump from one to the next without touching the ground. That may sound inconsequential; however, it becomes exceedingly important when a hunter is stalking his prey. Avoiding the ground, covered in oak and hickory leaves that can broadcast an approach from several hundred yards, can make the difference between bagging a squirrel or not.

I did a lot of hunting on Rocky Ridge and loved the quiet seclusion of the deep forest, the fall leaf color and the sounds of squirrels barking in the distance. I spent many hours in quiet solitude—whether or not I bagged game. I also hunted the area near the old chapel, along the Flat River, Boo-Hoo, and the swampland on Little River. They were all excellent areas for squirrel. Turkeys were not in abundance, but we did some turkey hunting in most of these areas. We hunted quail in the field areas where broom sage and young pines adjoined. One of my duties was to take the tractor, a bush and bog harrow and a sack of lespedeza seed into selected areas of the sage fields, cut out a swath and sow the seed in the mid-winter for next fall's bird hunting. The quail fed on the ripened lespedeza and we fed on the quail. While there were deer on the farm, they were so few that we rarely saw one. I cannot recall anyone ever killing a deer the entire ten years we lived there. 'Coon and 'possum hunting was great sport, as was an occasional foxhunt. This was done mostly at night and the thrill was to hear the hounds run the game, which sounded like music of a symphony.

Squirrel Hunting

Among my many hunting experiences, I recall one method of squirrel hunting that required not a gun but firecrackers. Any agile boy with a climbing rope, small firecrackers, matches and a burlap bag could profit from this exercise. First, find a squirrel hole or knot hole that looked like a well-used den, shimmy up the tree, light a firecracker, and throw it in the hole and slap a tow sack over the opening. When the "cracker" explodes, if a squirrel is in the nest, the hunter has bagged a mad squirrel! My great uncle Walter taught me another little trick of getting squirrel without a gun. Walk along a creek bank until the hunter finds a den at the base of an overhanging tree, cut a dogwood limb six or more feet long and leave it with a three-prong tip at the small end. Run the switch up the tree trunk inside until a soft resistance is felt. Start twisting the switch until the tail is wound around the forks, and keep twisting as the dogwood is slowly withdrawn. To affect the exercise in safety, a pair of thick work gloves is necessary. When that squirrel is pulled out, he is all fur and a few ounces of fury, and they have sharp teeth! Both systems work—even if I haven't gotten anyone to believe it.

Fishing

I spoke earlier of the flooding of the Flat River bottomland and the devastation it could cause. While bringing rich, fertile soil to river bottoms, it also created a fish "jubilee." As the water receded, dozens of small pockets of water were left full of bream and shell crackers to be picked up by the bucketful. While this wasn't common, it did happen on occasion, and we took full advantage of the condition. In fact, that was about the extent of my fishing at Fairntosh.

Oneness

It is difficult to understand what it is like to hunt in a deep, beautiful forest unless you have experienced it. The modern anti-hunters rave on about the "blood" hunter, those who are insensitive to the welfare of the animals. Well, we hunters are the ones who have provided the funds to bring numerous species back from extinction to abundance. Ducks, deer, turkey, mountain sheep, and goats, and on and on - that is the legacy of the hunter who cares. Of course, some don't care, but they are in a small minority. To be in the forest on a cool October morning, listening to God's world coming to life, is to know what the beauty of life is all about. It is a oneness with the Almighty.

For the Birds

As a small fellow, I was assigned the task of feeding and watering the chickens, gathering the eggs, and keeping a supply of wood for the stove and fireplace on the back porch. In those years, we had "Domineckers" to supply our eggs and fryers. Mother sold eggs, butter, and cheese to make some extra spending money, so eggs were very important as a source of income. One of our hens was named Jane, and she was a family pet. We would call her name, and she would come running out of the flock to get a special morsel or just to be petted. We also had a rooster that was aggressively mean; he would lower his head, fluff his feathers, and come charging at any intruder if his territory was invaded. At about the age of ten, I was tired of fighting Mr. Rooster and took a tobacco stick into the chicken lot with me one day. When he came to flog me, I whacked him over the head with the stick and broke his neck. I was scared out of my wits—what would Dad say when he found out that I killed one of the chickens? I promptly carried Mr. Rooster into the woods behind the chicken lot, dug a deep grave, and buried him. Sometime later, a few days perhaps, Mother lamented that she had missed Mr. Rooster but was relieved that he had disappeared. I never told what happened.

We also had White Holland turkeys and raised quite a few of them, although they were difficult to raise. In the springtime, when the poults would begin to grow, they would squat in a spring shower, turn their heads upward, open their mouths, and drown! Nevertheless, some did survive. One old gobbler that must have weighed 35 to 40 pounds was so clumsy that he had a hard time going to roost at night. Each evening about sundown, we would all gather on the back porch, using the steps as bleachers, and prepare for the show that we knew was coming. The gobbler would get back in the garden adjacent to the chicken lot, survey a big red oak tree in the chicken lot, and get a running start, flapping his wings all the while. Inevitably, he would realize the he did not have enough "steam" to clear the six-foot fence between the garden and the chicken lot and would "retrofire." On his second try, he would back off maybe another thirty or forty feet and try again with the same result. Maybe the third or fourth try, he would clear the fence, climb and circle to his right into the oak tree. When he landed, he would emit a triumphant gobble. We could then go back into the house - the show was over.

Two of the gobblers got into a fight one morning near the pump house. They were establishing territorial rights, I suppose. I had a gray kitten that insisted on placing himself between the fighting gobblers. That so distracted them that they finally gave up and quit. The cat won.

Lay of the Land

In addition to the Big House and the manager's quarters, numerous outbuildings were located on the lawn. The lawn itself was spacious and contained a variety of oak, elm, holly, and cedar trees. The oaks, both red and white, were enormous and graceful. The mansion faced west toward the old Richmond to Raleigh Pike. The pike can still be seen along the western boundary of the lawn, depressed five to six feet and perhaps thirty feet wide. The lane that connected the Big House lawn to the Old Oxford Road, and had replaced the pike some 100 years before we moved there, was lined with massive cedars of 40 feet or more. In the nearly 50 years since I left, the cedars have all died and were replaced by elms, just as pretty as the cedar. The lane itself was a half-mile long.

The outbuildings included the old Martini House (goodness knows where that name came from), a schoolroom, and the office of the mansion. On the other side was a line of buildings in a row, beginning with the kitchen, followed by the oil house, granary, smokehouse, and a small house for the domestic help. Behind the house was another well with a brick cooler room. Down in the formal garden, or it could have been a vegetable garden, was the outhouse. It must have contained six holes—obviously for the domestic help. Back in the woods, perhaps a quarter of a mile north of the lawn was the chapel and graveyard. All are still standing. A rock wall in back divided the mansion grounds and gardens from the slave quarters. Three or perhaps four dwellings were located behind the wall. There was also a huge scuppernong vineyard of approximately one acre and a smaller one of muscadines. At the south end of the garden were a harness shop and carriage shed. Then, farther south were the big barn and the long barn. The big barn had a cupola, or steeple, on the crown, where Colonel Cameron, a former owner of Fairntosh, sat to watch his race horses work out on a track he had constructed in the lower meadow. The track was still in evidence when we lived at Fairntosh, although half of it was overgrown with young pines. While we lived there, Dad had a long tool shed/granary constructed, as well as a corncrib.

Other sections of the plantation contained long rows of slave quarters, barns, and tool sheds. Eno Quarters had four or more two-story, four roomed houses. Stagville had even more, as did Horton Grove. The slave quarters were all of the same construction: two rooms down and two up, with stairs in the middle, and a fireplace on each end that warmed all rooms. The outside was pinewood, unpainted vertical rough board, while the inside walls were dressed pine. The space between the outer and inner walls was "chinked" with handmade brick for insulation. The quarters were quite sturdy and several remain today, after nearly two hundred years of wear and weather. At Horton Grove, a large barn of hand-hewn pine timbers was the center of attraction, and still is. The massive beams were held in place with pegs in an intricate design that would do credit to any modern architect. The barn still stands and is in excellent repair, having been donated to the North Carolina State Historical Preservation Society along with the Stagville home and several remaining slave quarters. All are being, or have been, repaired and maintained in good order. The Stagville center is a workshop demonstrating early building construction in North Carolina and is well attended by historians and architects alike.

PLANTATION LIVING

During the early 30's, 40 tenant families were living on the plantation, many of whom were merely there because of free quarters. After a couple of years, Dad pruned the tenant population in half and left only those who farmed the land and provided an income for the plantation. It was a tough job. Before moving to Fairntosh when asked what he intended to raise, Dad would reply, "cane." And raise "cane" he did! First, he had to establish authority. That included whipping a couple of the rough blacks that questioned his word. Then he proved to be as much a man as any of the hired hands by showing his great strength in lifting contests and other manly feats that the boys could understand. Within a short period, they began calling him "Cap," and his authority was established.

Tenant Living

Dad held court over neighbor fights, domestic quarrels, stabbings, shootings, and whatever else was presented. Unless a serious crime was committed, those incidents at Fairntosh never reached the courts. On one such occasion, two young blacks went squirrel hunting with a .22 rifle. One had gotten the other's sister in a family way and refused to honor the contract. They came back that afternoon without a squirrel, but with a .22 slug wound in the offending man's arm—the couple married post-haste.

Another time, one of the tenants got angry with his wife because she wanted to visit her parents on another section of the plantation. She insisted she was going, and he threatened to shoot her if she did. As she left in a huff, he took his old single barrel 12 gauge and blasted her when she got a hundred yards or so down the road. Having had his fun, and doing nothing more than scaring her, he went back in the house and began eating his supper beans and cornbread. Now, she was a little woman, a hundred pounds soaking wet, and he was a giant of six feet six inches, but she was mad! She slipped up behind him with a potato digger in hand and began cultivating his back. Then she proceeded to visit her folks. The next day, he came to Dad and pled his case to get his wife to come back home. Obviously, Dad was to be the intermediary. When asked why he shot at her, he replied, "Law Cap Taylor, I was jes funnin' her." Dad intervened, and the couple made up, but I doubt he ever shot at or around her again.

We also had several white tenants, perhaps five or six families that lived on the place. While they and the blacks usually were at peace, occasionally a rift would develop between them. A white man by the name of Hackney, and a black man, whose name was Gus Haskins, got in Hackney's Model T truck and went after a gallon of "Moonshine" on a cold, sub-zero night. On their return trip, a mile from Hackney's home, the truck skidded off the road – with the help of an over indulgence of "corn squeezin's" – and into a ditch. About eight inches of snow lay on the ground, and the temperature was five or six below zero, but that didn't bother the well-oiled pair. They set off on foot, and Haskins informed Hackney that he was going to spend the night with him. "Not so," said Hackney. They were close enough friends to drink from the same jug, but not close enough to sleep together. A fight ensued, which Haskins won. He proceeded to beat Hackney into unconsciousness with a guitar, poured kerosene over him from the lantern that they carried, and set him on fire. Fortunately, the poor fellow was covered with snow, so the fire quickly died with little consequence. However, Hackney lay in the sub-zero weather until after daylight when some of the neighbors heard his cries for help and rescued him. As a result of his exposure, Hackney lost all his fingers and toes to frostbite. Afterward, he was known as "Nub's" Hackney.

Raising Real Cain

Haskins, who was a nomad, took off into the woods with his pack of hounds and disappeared. A posse was formed, and Gus was hunted for several days until the posse finally gave up. A week or so later, Gus was cornered in the northern part of the county, still with his hounds. He did surrender, was jailed, tried, and convicted of attempted murder. He served a year of his ten-year sentence and died. Some said he couldn't take the confinement.

Still another case involved cousins who were vying for the favors of a young girl. They got into a fracas, and one shot the other in the mouth but didn't kill him. Oddly, the bullet did little damage. The offended sued his cousin for attempted murder, but before the case came to trial, they had settled their quarrel. When they got to the witness stand, one claimed that they were playing with the gun, and it accidentally discharged while the other claimed that he was cleaning the gun, and it accidentally discharged. The judge threw them both in jail for 30 days.

Two of our hands got into a fight over the length of hay bales that were being fed through the baler. The men, one white and the other black, were both well-muscled and about equal in size. However, the white hit the black over the head with a metal brake-rod and drove him to his knees. At one point, they were trying to feed each other into the mouth of the baler with the thing still running. Finally, the black broke away and went to the house where he was staying and got a shotgun to kill his opponent. By this time, Dad had been notified of the scuffle and intercepted the gun totin' man before he could do harm to anyone. Dad took the gun, calmed the man, took him to the office to draw his pay and sent him packing. He probably saved a life that day.

A Collection of Characters

Many characters lived on the plantation, and I won't try to describe them individually; however, a few are worth mentioning. Uncle Doc Edwards and Uncle Mose Shaw were both freedmen, and each claimed to be over 100 years of age. Since Mose was a plowboy and Doc was a water boy during the Civil War, it was assumed that Mose was the older of the two. He was a Primitive Baptist Elder at a local church while Doc bootlegged for a living. Neither had struck a lick at work for years, but since they were two of the original slaves, no one cared. Besides, they were on up in years and really too old to do much.

Aunt Judy Haskins, daughter of Uncle Doc and wife of the aforementioned Gus, was the lady who helped raise me and spoil me as much as possible. She was a kind, gentle soul who was as dependable as they come. In addition to doing washing and housework for Mother, Judy also kept the Big House clean and aired. I enjoyed playing tricks on her because she was so gullible. One morning before leaving for school, I stuffed my pillow inside my work shirt and put the shirt inside my overalls and stretched the whole thing out on my bed. I stuck my boots inside the breeches legs to fill them up and placed my hat over the spot where my head would have been. When Judy went upstairs to make the beds, as was her daily custom, she saw what she thought was me stretched out on the bed. She came screaming back downstairs, "Lawd God Miss

Taylor, Glenn's up there dead!" Mother went back upstairs with her and convinced her that I had, in fact, gone to school. We all got a chuckle out of the incident—even Judy. She would take special care in ironing my clothes as she knew exactly how I liked them. I looked to Judy almost as a second mother. While she was good around the house, she was a horrible cook. No wonder Gus took off with his hounds and lived in the woods.

The Worst of Times; The Best of Times

Life was not simple, but we had it much better than most farmers during the depression years. Dad drew a monthly salary, plus a bonus if the farm prospered, and he had a small V.A. pension from a lung condition contracted in France during World War I. Moreover, our housing was free—no heat or light bills to pay. All our beef, pork, and poultry were supplied by the farm, as was milk, butter, and vegetables, all of which we had to produce. Our clothing needs were simple: a Sunday outfit, school and work clothes, two pairs of shoes—dress and work—as was the style. Actually, we were much better off, in most cases, that our neighbors. They depended so much on their tobacco crops for their spending money to buy the necessities of life that if they had a bad crop year, they were in trouble. By no means was the poverty universal, however; the community did have many prosperous farmers.

When I was eight or nine, my parents took me to town to get a school outfit. I got an aviator's cap, complete with goggles, a sheepskin jacket, a pair of corduroy knickers, and high-top leather lace-up boots with a Boy Scout knife in a pocket attached to the left boot. To top it all off, I had a pair of Roy Rogers gloves with fringe along the cuff. Now, who can deny that I was the best dressed "dougan" in the third grade? I remember quite well that some of the kids came to school barefooted almost until Christmas and didn't come back until March. They had neither jacket nor sweater. One such lad was "Billygoat" Crabtree. He was a tough little cuss, but fat and muscular. I don't think he was afraid of the devil himself. His cousin, Arthur, was 14 and still in the third grade. I doubt that either made it to the fourth, but they could sign their names to a tobacco warehouse bill and knew how to grow tobacco by the time they could walk. Our school bus was loaded with an assortment of kids from the first grade through high school. Some poor, some not so poor, but all were bound by a common bond—need.

Weldon Hester, a schoolmate, and his brother were none too bright, but both attended school at Bahama. One Monday morning, both got on the bus with their right index finger bandaged. They were reluctant to tell what happened, but it finally surfaced that they had seen a neighbor who had a missing finger and began to wonder what it would be like to have a missing digit. They went to the chopping block in back of the house, and while Weldon, who was the older, held his finger on the block, his little brother took a sharp axe and chopped it off. With the first spurt of blood, kid brother got cold feet and ran; however, Weldon, being swifter, ran him down, brought him back to the block and chopped his finger off. At least they didn't have to write in school for the next few weeks, and who knows, that may have been the whole point of the exercise.

The bus ride was an experience within itself. Ours was an old dilapidated 1925 International that had the headlights up beside the windshield instead of between the fender and radiator. It was of wooden body construction and was so long that it was tail heavy. To compensate for the uneven weight distribution, the county garage had installed a second set of wheels in tandem behind the rear wheels. We soon discovered that by all the kids rushing to the back of the bus, we could tilt the front end off the ground, causing the driver to completely lose control of the vehicle and wind up in a side ditch. The old bus would not be going much over 20 miles per hour, so there little danger to the riders. We could either await another bus or walk the two or three remaining miles to school. We did a lot of walking back then.

LEARNING ABOUT LIFE

Many other people lived at Fairntosh—among whom were the "hired hands" who became almost a part of my family—both black and white. Emmit Tilley, our first herdsman, was a big strapping fellow with a high-pitched voice. He was strong and took no truck from any of the other boys. Whether winter or summer, Emmit always wore an extra pair of pants under his overalls. It was whispered, and never questioned, that he carried a small revolver in his inside pants pocket. He always carried a pair of hooknose pliers in the leg pocket of his overalls and would grab the big barn rats by the tail with those pliers as they ran up and over the tack closets in the barn hallway. Apparently, once they got out of sight, the rats thought they were safe; however, they often would leave their tails hanging over the top of the tack, and that was Emmit's cue. We had more bobtail rats running around the stables than those with tails.

Mentors

Emmit was my first mentor. He took me under his wing when I was a little boy of seven and looked after me and taught me much, including profanity. Our hands were basically a good group of men with little education and a propensity for profanity. I learned all the "no-no" words at a very tender age and could cuss with the biggest of them. Emmit and his wife, Claire, lived in a small, two room shack behind the wall as long as he was herdsman. After he left the farm, Joe Hart took the job of herdsman. Like Emmit, Joe wore rubber knee boots winter and summer. We often marveled at their resistance to foot disease; however, cattle pens are notorious for their black quagmire, and I suppose that wearing boots was the only answer.

Others that I recall were Jerry Peak, James Wallace Justice, Collin

Haskins (Aunt Judy's son), Norfolk Burton, Paul Taylor (no kin), and many others who were seasonal workers. The white hands included Wilbur Taylor (my cousin), Coy and Fred Luxton (brothers), and others that were seasonal. During harvest season (grain, hay, corn, etc.) we would employ 15 or more hands, but during the winter months, perhaps a half dozen regulars. Winter duties included plowing, weather and soil permitting, tending livestock, shoeing the horses and mules, mending harness, repairing buildings, grinding feed, fixing machinery, and a thousand little jobs that were necessary to keep a large farm going.

Timing is Everything

Considerable effort was spent on getting the tobacco to market in the fall, which entailed picking up the different lots from the many farmers who raised tobacco at Fairntosh, taking the farmer and his tobacco to market, making sure that the farm got its share of the sale, and returning the farmer home, usually loaded with both cash and booze. That was the one time of the year they had money enough to buy frivolous things like liquor. Word would go out on certain days that the truck would be available to haul tobacco to market the following day, and those who had graded and tied their tobacco would let it be known that they wanted their crop hauled to the Durham market. Tobacco was the only cash crop grown by the tenants. All other crops—corn, grain, and meat – were grown for consumption by the tenant family. Of course, the farm got its share of that also.

Fairntosh had two distinct farm operations: one managed by the tenants, and the other by the farm itself. The farm was primarily involved in raising Angus cattle for registered sale and grade stock for beef. We ran somewhere between 125 and 150 head of registered and grade beef cattle. At any given time, the registered Angus population would be about one third to one half the total number. The herd, coupled with the 12 to 15 head of horses and mules plus sheep, goats, hogs, and poultry, meant that an enormous amount of feed was required to keep the livestock well fed. Everything had to be done on a seasonal basis and began in the late fall with the harvest of the corn crop, turning the land from spring grain and sowing the grain, all by Thanksgiving. By February, we top-dressed the spring grain with nitrate of soda and hand broadcast Korean lespedeza as a second crop on the same land. Then by April, we began planting the corn crop, followed by harvesting the small grain in June and early July. We baled most of our straw for cattle and horse bedding. The haying season followed in July and August. We filled three barns with lespedeza hay, both in bulk and bales, and this was perhaps the dustiest, dirtiest job of them all. Following hay season closely was cutting and grinding the silage for two silos to feed the cattle green foodstuff through the winter months. The process was never ending. While the corn grew, it had to be cultivated several times during the spring months until it was all "laid by."

THE JOYS OF HARVEST

We had two tractors by the late 1930's, and while one harvested grain, the other would cultivate corn. In addition to the F-20 Farmall, we also bought an "M" Farmall for the heavy-duty work. About this time, after the loss of a 17-acre field of barley in a 1938 flood, Dad bought a combine. I was so intrigued by the combine that I was accused of sleeping with it. More importantly, I learned every nut, bolt, chain, belt, and piece of steel in the whole thing. I was assigned to bag the grain and repair whatever went wrong. That had been my duty during the days of the old threshing machine – to sit on the tractor and cut the power at the first sign of trouble, and then climb inside the hot bowels of the machine and fix it. During July threshing, it could get 120 degrees inside one of those pesky things, but I thought my job was the most important one of all—little realizing that all the other boys stood back with the attitude: "Well, if he's fool enough to go in there…"

By the second year of the combine harvest, I had graduated to tractor driver and mechanic. I suppose I did have an aptitude for fixing things, because on the farm everyone has to be resourceful. There isn't a garage around the corner to call upon at every crisis. I drove and Joe Hart or James Wallace Justice bagged. We usually began combining around 8:30 or nine o'clock in the morning and threshed until evening when the dew began to fall and cause the grain to get "tough." It was hard work, but rewarding, and I enjoyed grain harvest above all other duties on the farm, except handling cattle. Joe or James Wallace and I would get to the fields by 7:30 A.M. with a full tank in the tractor, oil and grease the equipment, adjust the belts, and make any necessary repairs or adjustments in time to start threshing as soon as practical. The tractor never stopped except for fuel, water, or a quick relief trip all day long. At lunchtime, Dad would bring someone with him to relieve us while we ate.

Those were some of the more memorable meals I can remember on the farm. Mother wouldn't let me take a pack lunch like the rest of the field hands; she sent a hot meal of cornbread, snap beans, tomatoes, cobbler pie, and a half-gallon thermos of fresh, cold whole milk. Of course, I was hungry enough to eat the tires off the tractor!

Asleep at the Wheel

One day, Joe and I were combining the wet field, across a large drainage ditch from the barn field where V.C. was cultivating corn. He had gotten caught coming in the house after daylight that Monday morning, and Dad had sent him straight in the field after breakfast, without a wink of sleep. On up in the morning, I noticed him nodding as he sat on the old F-20, cultivating his corn, and I wondered how many plants he was plowing up as he went. The ditch, which was at least 15 feet deep and nearly as wide, was at one end of the rows and the river was at the other end. It was only a matter of time until something happened.

At some point, our combine choked down with vetch or something, and we cut off the tractor to correct the problem. There was an unnatural quiet. I didn't hear V.C.'s tractor in the other field, nor did I see it when I stood up on the seat of my tractor. Both Joe and I ran over to the barn field to see whether he had gone in the river or the drain ditch—it was the ditch! Belly up, the tractor was in the ditch with the cultivator prongs sticking up like a tiger ready to attack. V.C. was climbing up the bank unharmed as we approached. As if on signal, while we stood there discussing his predicament, we saw a cloud of dust approach, kicked up by Dad's little green Chevrolet pickup. I started to leave in panic, but V.C. prevailed upon me to stay for moral support. As Dad came boiling out of the truck, seeing that everybody had survived, he shouted, "How could you be so stupid as to back a tractor into a ditch like that?" To which V.C. replied; "Back it, Hell, I drove it in head first!" It took two days to rig up an "A" frame and with the help of a chain hoist, everyone available, and the other tractor, we were able to extract it from the ditch with amazingly little damage.

Labor of Love

In our earlier days, during the depths of the depression, labor was so cheap that it was almost like slavery. Men worked for a mere 50 cents a day—and were glad to get it. In those days, living at Fairntosh was almost like living in the pre-Civil War period. The black people took off their hat when addressing a white, always came to the back door (it was considered "uppity" to do otherwise) and addressed whites with "yessa" or "nawsa." Dad hired a dozen or more blacks to hoe the corn crop. The farm grew upward of 100 acres of corn and that meant a lot of hoeing. The "wet-field" rows were a half mile long and laid off in lands, maybe 50 to 75 rows per land. The hands would be "chopping" away and Emmit, who was straw boss then, would stop, lean on his hoe, and sing a refrain leading to an old Negro spiritual. In response and in harmony, the gang would catch the tune. As they hoed, they sang some of the most beautiful and melodious music I have ever been privileged to hear. That only lasted a season or two because the pay scale rose, and farming became more sophisticated. I wouldn't have missed that short peek at the past for the world.

I am reminded of the second generation of blacks at Fairntosh, who were all tenants: Sy Hart, Hirman Burton, Amos Sowell, and others whose faces are memorable, but their names forgotten in the ensuing 50 or 60 years. Among the whites were Willie Ellis, Steve Eades, Arthur Ashburn, Edgar Williams, an elderly gentleman by the name of Stone, and another whose last name was O'Brien. At Eno Quarters were Ira Veasey and Ike Parrish, both of whom had been devastated by the great depression, but were good farmers.

Some of these, through husbandry of resources and soil, were able to purchase their own farms after WWII as a result of high crop prices and low land prices in the mid 40's. Several of them had their own tractors and trucks, as well as a host of other farm implements, while still tenant farming at Fairntosh. Others, who were less imaginative or industrious, remained as tenants on the farm until it was sold.

GROWING OLDER

On Fairntosh during the 1930's, we had neither electricity nor central heat. Our house was old and drafty with the front part of the house always cold in the winter and hot in the summer, but we did have amenities that made life more bearable. In addition to our water system and bathroom, we had an Electrolux kerosene refrigerator, a Silverstone battery operated radio, Aladdin kerosene lamps, a roaring fire in the dining room fireplace, and ample stove wood for Mother's kitchen range. Mother canned fruit and vegetables in great quantities, as well as sausage. We made gallons of muscadine, scuppernong, and blackberry wines. Dad experimented with a batch of brandied peaches, which had to be buried in the ground to age a year or so, but that didn't work. The jars exploded on the pantry shelf once they were taken out of the ground. Hard liquor was never allowed in the house until I had finished high school and was ready for college.

I Wanted a Bicycle but Got a Pony

On my tenth birthday, Dad gave us an Indian pinto pony named "Lady." According to the trade stable owner, she was a sheep pony from Wyoming. While both V.C. and I claimed ownership, I always knew in my heart that Dad gave her to me. He wouldn't allow me to have a bicycle due to my asthmatic condition in my early years, even though I always longed for one. My schoolmates, who had bicycles, all wanted ponies. I suppose there is a lesson in there somewhere.

Lady was our pride and joy. We seldom rode her with a saddle, and often without a bridle, just a plow-line fashioned into a halter. We guided her by knee pressure or neck pressure. She refused to work in harness and would rare up and fall over backward in the wagon shaft. On one such occasion, the boys had hooked her up to the one-horse wagon and were attempting to haul some corn to the hammermill for grinding into feed. As usual, she balked. They backed the tractor alongside her and wrapped a wire around a sparkplug, holding the wire with a fertilizer sack for insulation and stuck the other end of the wire under her tail. Unfortunately, she was pointed in the direction of a fence that divided the carriage house where the mill room was located at that time and the scuppernong vineyard. With the first volt, she lunged wildly through the fence and vineyard, and virtually destroyed the wagon, distributing the load of corn over a half-acre. Another one of her little "cuties" was a stiff-leg stop in full run, dropping her head and sending the rider into low orbit. You could never tell when she was going to pull one of her cute little tricks.

We traded Lady for a larger horse when I was 13 or 14. Her name was Bessie, and she was dependable and docile. We used her for riding, even though she was not gaited, and she was a good plow horse for the garden and light work around the barn. She was also a brood-mare that delivered at least two colts, Legs and Gray Boy. Gray Boy was my pet. I would call him in the pasture and feed him lump sugar, until he realized that the sugar came from my shirt pocket. Then he would grab the pocket with his teeth, rip it off, and help himself to the sugar cubes on the ground.

Following my experiences with Bessie, I graduated to Nancy. Now that was a horse! She could rack, trot, single-foot, run, and walk, so I suppose she could be considered five-gaited. She was also a good herd horse and could be guided by the rider's weight in the saddle, knee pressure, or with the reins. She could cut a 90-degree turn as clean as any horse I have ever seen, most important in running down a stray steer. I spent long hours riding over the plantation for the sheer pleasure of riding horseback. We would go hunting up in the vast forests of Boo Hoo.

Nancy was extremely intelligent also. She didn't like to have her ears touched, but would allow the rider to take a small brush and "swish" sheep flies off them while traveling through the woods. Apparently, she disliked sheep flies more. On one occasion, I was herding cattle in the low ground, next to the riverbank, when one of the cows broke and ran across a "slew" that emptied into the river. Determined to cut off the cow, I reined Nancy to the "slew" path and started down the bank after the cow. "Not so," said Nancy! She would not, absolutely would not, go down that bank into the cut. I tried to pull her, tap her with the rein end, and kick her in the flanks, curse— all to no avail. In desperation, I dismounted and tried to lead her into the cut, only to have my feet shoot out from under me in the slick, muddy bank. Somehow she knew the path and cut were too slippery for her to negotiate, and she simply refused to be misled.

She was accustomed to the noise of a "bull whip" cracking around her as we used the whip's noise to herd cattle; however, when I tried to fire a 20 gauge shotgun at a woodcock from the saddle, she literally squatted and jumped out from under me, leaving me astride thin air. I hit the ground, missed the bird, and she ran off a few feet and stopped. As I approached her, gun still in hand, she shied away and refused to allow me to approach. I laid the gun on the ground and walked peacefully over to her and took her by the reins. I led her off through the woods a hundred yards or so and tied her to the limb of a tree. I then retreated to the gun, unloaded it, and placed it head-high on a dogwood limb. I headed back to Nancy, mounted and rode back under the gun and retrieved it on the go. She never knew when I had retrieved the shotgun.

Life Moves On...and So Did We

Our last ride was almost fatal for the both of us. By the fall of 1941, I was a student at North Carolina State and could only come home to hunt and ride on weekends, usually Saturday and Sunday afternoons. This left Nancy plenty of time to stand in the stall and build up excessive energy. Some weekends we didn't go out at all. That December, I had finished the fall semester at State. It was the last few days or weeks before we moved from Fairntosh to Durham. I decided to go squirrel hunting in the Boo Hoo woods and rode Nancy into one of my favorite hunting spots near the location I had tried to shoot a woodcock a year or so before. I hunted for a couple of hours, but for some reason, the squirrels were not moving.

In the late afternoon, I gave up, empty handed, mounted the horse with a .22 rifle stuck in my belt, and started home. She was frisky and wanted to run. I gave her her head for a mile or so, but when I tried to rein her in, I simply could not do it. She had her head and was determined to keep her pace all the way to the barn. We were running on the shoulder of the highway, which runs perpendicular to the lane into the mansion grounds. I still wasn't worried because I enjoyed feeling her strength and speed, and even though I couldn't stop her, I erroneously rationalized that I could just let her run on down the highway past the lane, until she got it out of her system.

That might have worked, except she wanted to go to the barn. We had a conflict of will—she was heading for the barn, and I was just as determined to continue down the highway. Obviously, we couldn't do both! I should have let her have her way. Instead, I pulled the right rein until I had her head turned sideways, throwing her in the middle of the tar and gravel road at the end of the lane. I don't know how many thoughts flashed through my mind as I felt her going down on the pavement, but I do recall that I expected to be killed! We hit with a bang, and she wound up thirty feet from me and the spot where we fell. I was about that far from the same spot. I regained consciousness about the same time she started to get up, amazed that I was still alive, and even more amazed that I had no broken bones. However, I left skin all over the place.

My kneecap was badly torn, as were my ankle and elbow on my left side. My shirt, pants, and boots were in shreds as was the rifle. Nancy had a small patch of hair about the size of a silver dollar burned off her left shoulder. She ran on to the barn and Joe Hart, who was one of Dad's herdsmen, saw her come in riderless. He climbed aboard Nancy and came looking for me. We met as I had limped a few hundred yards down the lane. Having determined that I was battered but not broken, Joe rode back to the barn and waited for me with a bottle of blue gall lotion, which was used to doctor the livestock's harness sores. His administering that stuff made my raw wounds hurt more than the actual fall. However, my patches healed well after the gall lotion treatment. As a result of the fall, I had a layer of tar under the scar that lasted for several years. That black knee scar almost kept me out of the army a year and a half later.

I have fond memories of Nancy and of my great joy in riding her to the distant parts of the farm, herding cattle, milking her when her week-old colt died, and incidents too numerous to mention. Nancy was a great lady.

HAULING CATTLE

Cattle business was big business at Fairntosh. In the late '30s's we could sell a registered Angus for upward of two hundred dollars at birth. The grade stock sold at regular beef market prices and was marketed locally or on the livestock market in Richmond. Dad's training was ideal for the cattle business since he had his degree in Animal Husbandry. Shortly after his arrival at Fairntosh, he began looking at both Herefords and Angus, deciding on the latter as the registered stock and a cross of Angus on Hereford for beef. His reasoning centered on the Angus' heartiness, and the Hereford's docility. He deducted that a cross would range better, herd better, and thereby produce better beef. He would not allow salt licks. He reasoned that by licking salt, a cow would wear off the roughness of the tongue, and thereby get less grain in a "sweep" of the tongue while grazing. A ten percent loss of grazing would materially affect the weight gain of a cow. Cattle only put on weight while they are lying down and chewing their cud, not while they are grazing, he further rationalized. It was all based on observation, rationale, science, and I suppose, discussion with the cows.

Cowboy Glenn

We had several large pastures of hundreds of acres and one on the Flat River of over a thousand. Additionally, there were two barn pastures of a hundred acres or so, a large pasture at Horton Grove and the Flat River or "low ground" pasture. Most of the summer ranging was done either in the river or Horton Grove pastures. The two barn pastures were for working cattle as well as for the horses and mules to graze. The 50 or more cattle grazing in the river pasture all summer would grow unaccustomed to people unless they were visited on a regular basis. One of my more pleasant assignments during those summer months was to take bags of table salt and ride Nancy to the pasture to "handle" the cattle once or twice a week. The object was to first find the herd in a large grazing area down along the river where they also drank their water, and place the salt in small clumps - double hand size - around in the grass and allow the cattle to lick it off the ground. I would stand in the midst of the herd and scratch them as I talked to them. They were completely docile, and I loved every minute of their rubbing and nudging me as they ate the salt.

The not-so-pleasant duties of cattle herding came at the time of flood along the river. We had an understanding with the operators of the dam up the river that when it reached flood stage, they would call us to give us time to get the cattle out of the low ground. Those phone calls usually came at midnight or later. Then the fun began. We'd go up to the barn, saddle up horses and mules and head out for the river in driving rain, thunder and lightning, trying to hunt down the herd in a vast pasture area. When they were located after a considerable search with lanterns and spotlights, we drove them to high ground in the holding pasture below the barn. I don't recall that we ever lost a single cow as a result of drowning. We didn't always find every cow or calf, but those left usually found high ground and were not harmed.

When we herded cattle from Horton Grove pasture to Fairntosh, a distance of a couple of miles, we would drive them down the highway. Anyone who had the misfortune of driving to town behind the herd had a long and messy wait. Cattle are not known for doing what the drivers want and often take off on their own. That's where a good cutting horse comes in, to chase them down, cut them off, and drive them back into the main herd. That is also where the "bullwhip" came in. It would make enough noise to frighten a cow into submission, most of the time.

I Think I'll be a Veterinarian…Maybe Not

I think it was the spring of 1940 when we discovered several of the yearling calves dying of Black-Leg. Black-Leg is diagnosed by a bleeding between the hide and flesh, and kills in a few short hours. At first, we thought one or two had been struck by lightning, but they kept dying. Our veterinarian diagnosed the disease and prescribed an inoculation. We spent the next three days vaccinating the entire herd, and the plague was over. All told, we lost eight calves, which represented a good portion of the previous year's calf crop, maybe a third. The vet and I were dissecting a calf one Sunday morning after it was found dead the day before. It had lain out in the hot sun for at least a day and was getting "sweet." The vet peeled back the hide, scraped some blood on his scalpel and stuck it under my nose with the question, "Does this smell sweet?" I had aspirations of becoming a vet up until that moment. I knew then that I couldn't cut it!

Cattle Expedition

In the spring of 1941, Dad and Alton Sutton, our Assistant County Agent, and I made a trip to Virginia and Maryland on a cattle-buying expedition. I had become somewhat knowledgeable in beef cattle through my 4-H work, and Dad wanted some companionship, I suppose. The three of us visited Brandywine Farms, Red Gate Farms, and Cold Saturday Farms, all of which had some of the better Angus bloodlines in Virginia and Maryland. His pick was the Red Gate herd and twelve or fifteen of their stock, two of which, were bulls. One of these was a yearling, and the other was an enormous bull named Donhead. He was a half brother to the International Grand Champion Dam at the Chicago Livestock Exposition. While he wasn't rated, he came from an impeccable bloodline, and weighed 2,373 pounds. He immediately became my pet, and would follow me around the pasture like a puppy and bawl when I closed the gate between us.

Cattle Runs

During the summer of '41, I made three cattle hauls from Millwood, Virginia to Fairntosh. Each load could accommodate six or more cows, depending on their size. The routing took us over the Blue Ridge to Paris, Culpepper, Farmville, Oxford and home. While the road, old U.S. 15, was bad enough – part mountains and narrow road without shoulders – the worst part was driving live weight. A truckload of cattle is tricky business, if the driver doesn't know what he's doing. When the cattle are first loaded, the cattle want to sway with the truck. After thirty miles or so, they begin to learn to lean into the curve and provide a counterbalance to the truck sway. If the driver knows when to slow down and when to accelerate into a curve, he can actually train the cattle to the speed and motion of the vehicle. On the three trips we made to Millwood, Joe Hart, V.C., and Clarence Duke were my companions. But, V.C. and I were the ones that had the real adventure. Coming into Culpepper with a full load, a bread truck got in the way as V.C. swung a 90-degree corner and had to cut hard right to avoid collision. Our truck tilted and paused on the left wheels for what seemed an eternity. I was on the right side and felt as if I were on top of Pike's Peak. Fortunately, the cattle had ridden long enough to become trained and scrambled to the right (up) side of the truck. Their weight brought the truck down on all wheels. Almost immediately, first and second gears stripped and left us with a truckload of cattle 150 miles from home with only the two lower gears in action, and no more than ten dollars in our pockets. We drove back to a stockyard and unloaded our cattle, then slowly ground our way back to Culpepper where the Chevrolet dealer fixed the old 1937 model truck, filled the tank with fuel, let us have a few extra quarts of oil, and a ten dollar bill, all on a phone conversation with Dad. That was a trusting soul!

Years later, I asked Dad, who was not liberal with his compliments, why he sent a 17 year old boy to haul cattle across the Blue Ridge Mountains. I allowed that he must not have thought much of the safety of his son. He replied, "You were the best driver I had." Both V.C. and I were given responsibilities far beyond our years and expected to perform as men. I honestly don't think age was ever a consideration for Dad. Perhaps it had something to do with the fact that he was left fatherless at a very young age and had to shoulder responsibilities to help keep Grandmother and the younger children from need. Looking back on those early days, I cannot remember a time that I didn't think I could handle a task that was assigned to me. Sometimes, I didn't want to do it, but that was beside the point, I did it.

Newspaper clipping of author's innovative 4-H project, introducing Jarvis Golden Prolific seed corn to State of North Carolina, 1940

Glenn Taylor Wins In Judging Contest

Bahama Boy Presented Holstein Calf For Taking Event At Local Fair

Glenn Taylor of Bahama, Route 1, yesterday won first prize in the livestock-judging contest conducted by vocational agriculture teachers and Assistant County Farm Agent J. A. Sutton at the American Legion Fair.

Taylor was awarded a purebred Holstein calf donated by Long Meadow Dairy in recognition of his accomplishment.

Joe Hopson, Durham, Route 3, won second place; Jack Wilder, Morrisville, Route 1, third; and Felix Crabtree, Bahama, fourth.

The contest was open to all vocational agriculture departments and to Senior 4-H Clubs. Dairy cattle, swine, and poultry were judged.

The Mangum School team coached by C. M. Allen took team honors. The Oak Grove team coached by A. E. Williams was second; Lowes Grove coached by J. M. Hagy third; and the Bethesda team coached by W. I. Smith fourth.

Successful Youngster

Glenn Taylor . . . grows certified seed.

Young Farmer Is First To Grow Certified Seed Corn

16-Year-Old Glenn Taylor Of Fairntosh Farm Has Also Fattened And Sold Beef Steers At More Than Average Market Price—Owns 150 Leghorn Hens

Glenn E. Taylor, 16, of Fairntosh Farm, of which his father V. C. Taylor is superintendent, is the first farmer in Durham County to produce certified seed corn of any kind during the last three years.

Young Taylor has three acres of certified Jarvis Golden Prolific seed corn under cultivation this season. He is a member of the North Carolina Crop Improvement Association and will be a senior at the Bragtown High School during the session of 1940-41.

For the past four years Glenn has been an active member of the Durham 4-H Club. During this period of time he has fattened and sold seven beef steers. One of his beef steers was sold in Richmond a few weeks ago at a premium price of $9.75 per hundred, which exceeded the top price for that day of 75 cents per hundred pounds.

Young Taylor has been a member of the judging teams at the Rocky Mount Stock Fair and at the State Fair three times. At present he has approximately $600 tied up in various farm projects in the 4-H Club program under the direction of Assistant Farm Agent J. A. Sutton.

At present Mr. Taylor has two Jersey cows, one pure-bred Berkshire gilt and 150 pure-bred Leghorn hens. His hens are valued at $225, and will begin laying within the next few weeks.

Besides these projects young Taylor is active in every phase of farm life. He was a member of the dairy demonstration team at North Carolina State College during the short course given there this Summer.

According to Assistant Farm Agent J. A. Sutton, Durham County bought approximately 100 bushels of certified seed corn last year. But with a good crop season Glenn Taylor will be in position to furnish the needs of many farmers wishing certified seed corn next Spring.

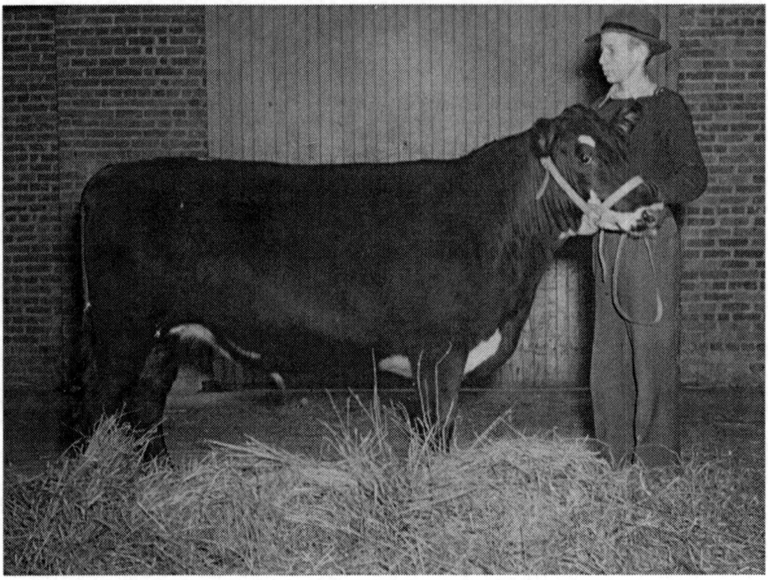

Author with steer in the Rocky Mount Fat Stock Show, 1940

Team representing North Carolina in the World Poultry Congress, Cleveland, Ohio in 1939. (L to R) : Alfred Green, Glenn Taylor, Roy Walker – all from Durham County NC

The manager's house at Fairntosh with '41 Dodge in front. Dodge was purchased by author's father a few days before December 7, 1941. Vehicle later became author's transportation while in college on G.I. Bill after WWII.

Author with 21 jewel Waltham watch awarded to him for winning NC State Champion 4-H Baby Beef Award, 1941

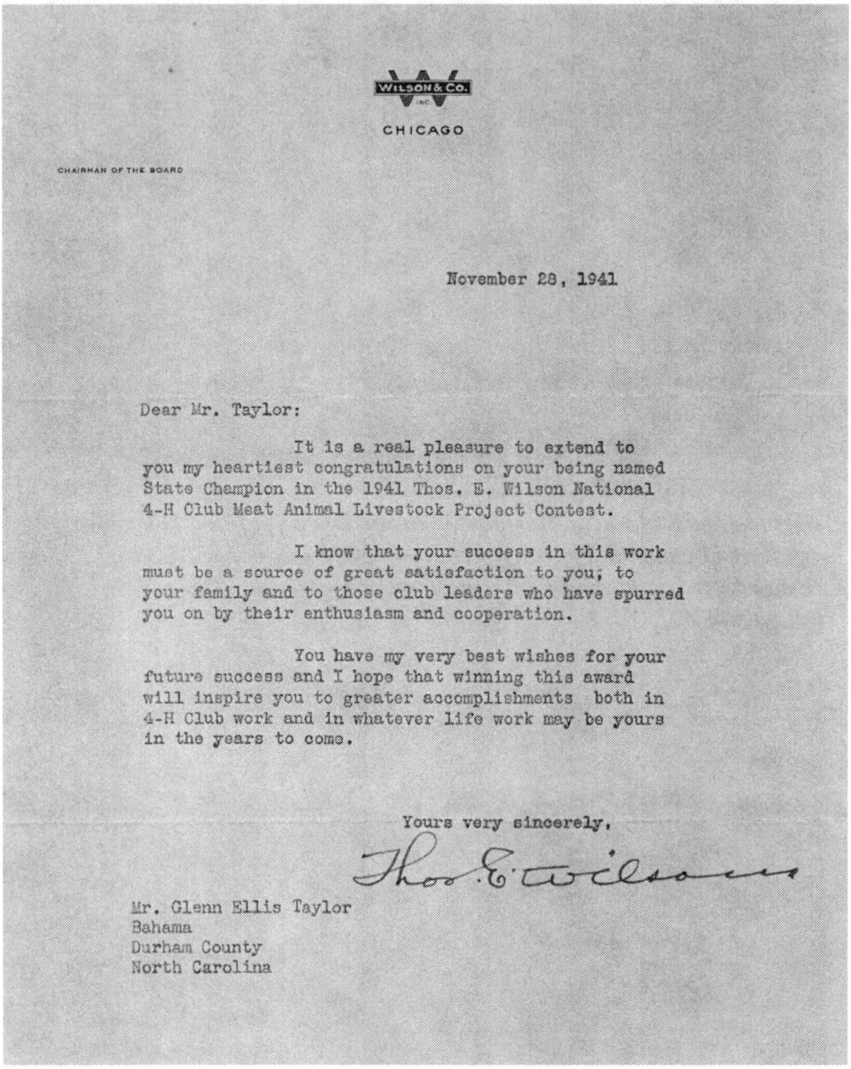

WILSON & CO. INC.

CHICAGO

CHAIRMAN OF THE BOARD

November 28, 1941

Dear Mr. Taylor:

It is a real pleasure to extend to you my heartiest congratulations on your being named State Champion in the 1941 Thos. E. Wilson National 4-H Club Meat Animal Livestock Project Contest.

I know that your success in this work must be a source of great satisfaction to you; to your family and to those club leaders who have spurred you on by their enthusiasm and cooperation.

You have my very best wishes for your future success and I hope that winning this award will inspire you to greater accomplishments both in 4-H Club work and in whatever life work may be yours in the years to come.

Yours very sincerely,

Thos. E. Wilson

Mr. Glenn Ellis Taylor
Bahama
Durham County
North Carolina

Letter of congratulations from Thomas E. Wilson of Wilson & Co. from Chicago, IL on author's NC State Champion 4-H Baby Beef Award, 1941

Recent photo of the kitchen at Fairntosh

The Big Barn at Fairntosh

Stagville House (Early 1800's) where author's family lived while renovating the Manager's House at Fairntosh.

Slave Quarters at Horton Grove

Horton Grove barn

Manager's quarters at Fairntosh

The "Big House" at Fairntosh

Half-mile lane leading from main highway to owner's complex at Fairntosh

TIME TO MAN UP

Along the lines of responsibility, Dad decided in 1935, when V.C. was 14 and I was 11, that we should earn our own way financially, so he assigned us a three acre tobacco crop. We raised tobacco for five seasons, and I hated every day of it! In later years, I established a philosophy that anything beats suckering tobacco. I still believe it! We were sharecroppers, which meant that we got to keep 50 percent of the profits after all expenditures were deducted. V.C. being the larger of the two of us got a greater percentage. As I recall, the summer-long backbreaking crop would net me in the neighborhood of $100. From that, I had to buy my clothes, provide my own spending money, and meet other needs not related to food and lodging.

Suckering Tobacco

Tobacco is a 13 month, 375 day a year, 25 hour-a-day occupation. It controls the farmer, not the other way around. Between blue mold, black shank, hail, horn worms, drought, wind or rain, the farmer just might make a little money—if he's lucky. Starting with the plant bed in February, tending it until time to transplant to tobacco slips in early May, through hoeing, worming, spraying, plowing, topping and suckering, priming, curing, grading and selling, tobacco is an all-consuming way of farming. Even when the farmer may think he has his crop in the barn and safely curing, disaster may still befall him. A loose leaf falling on a flue pipe or a bolt of lightning attracted to a hot tin roofed barn can ignite a fire that will destroy barn and contents in minutes. Tobacco, in those days, was flue cured. By flue cured, I mean that fires were lit in two furnaces and, through a series of pipes called flues, the heat was distributed over the dirt floor of the barn. The tobacco was hung green on a series of tier poles arranged four tiers high and maybe six across. The furnaces were set with a smoldering, low fire to yellow the tobacco before beginning to raise the heat. As the heat increases in slow stages, the sap is driven from the tobacco leaf to the fibers and then into the stem. When the stem is dry in a process known as "killing out" the tobacco, the heat in the barn ranges around 185 degrees. Then it is most susceptible to fire.

I suppose I must have had some memorable occasions linked to growing tobacco, but they are limited. Mostly, I remember the cold, rainy mornings we primed the plants (picking off the lower leaves that had yellowed), strung and hung the sticks of tobacco in the barn. Then we had the back-breaking job of suckering, or pinching off the growth buds, of the tobacco plant, and the all night stints at the barn tending furnaces. On the lighter side, I collected a couple of hundred Indian arrowheads, all perfect, while hoeing our first crops. The field was located on or near the permanent encampment of the Tuscarora tribe. I still have a couple of those arrowheads. In that particular field, bullbats circled endlessly eating moths that flew over the plants. When they would dive for their prey, they would emit a long, low "booonk." I could have watched them by the hour. Huckleberries, gooseberries, and plums grew around the field and provided a much needed respite during a long day's work.

BECOMING A WINNER

At about twelve or thirteen, I became involved in 4-H Club work. My mentor was Alton Sutton, the Assistant County Agent. I suppose he and Dad taught me everything I knew about animals, crops, and organization. At the risk of sounding boastful, I won about everything the 4-H Club of North Carolina had to offer. I started with raising and showing beef calves for the various fat stock shows and fairs throughout the state. I learned to feed, groom and show my steers to their best advantage. Knowing how wasn't enough to win ribbons, but doing was.

Winning is the Name of the Game

My first venture ended in disaster! We took three steers to the Rocky Mount Fat Stock Show in 1938. I knew I had a winner in my Angus steer, "Blackie." Enroute to the show, the truck caught fire from either a careless cigarette or straw sifting down to the muffler, and consumed my steers. We did salvage the meat by slaughtering the steers and selling the beef to a local butcher. Ironically, the name of the town was Nashville, NC, a town name that would have a place in my future, as I would spend years serving my railroad in Nashville, TN. But, as for the incident of the fire and destruction, I was devastated. I knew that my "Blackie" was the grand champion of the show—regardless of what the judges might have decided. I did participate in the fat stock judging contest, and won. The next year I showed my calves and while they didn't win the show prize, they did show well. I also made progress in my abilities to judge the beeves in the judging contest. After another year, I was ready for the big time. I showed and judged cattle in the county and state fairs, as well as the Rocky Mount Show. Winning was the game, and I won handily.

By the time I became a freshman at N.C. State, I was showing not only my calves, but the entire complement of show cattle at the farm. We split the winnings. That year, 1941, we took eight or ten cows to the State Fair in Raleigh for show. Donhead, our 2,300 pound Angus bull, won Reserve Champion and "Number Eleven" (one of our top cows) won Grand Champion Dam. While grooming Donhead for the arena, one of the "World of Mirth Show" executives came by my stall and became intrigued with the exercise. He sat on a bale of straw and watched as I washed Donhead in Lux soap, rubbed him with Vaseline hair oil, waved his flanks and shoulders to bring out his beef qualities, polished his hooves and nose ring, and curled his tail. I'll tell you, he was one handsome dude! Before I took him in the ring, the show executive commented, "fantastic," and reached in his black valise and proffered a handful of tickets to all the rides on the midway. As he was leaving, he turned back and said, "You will also need tickets to the shows," and gave me another handful of tickets. I had tickets to every ride and show on the midway, and only had to pay tax on the tickets I used.

That's the Ticket!

By virtue of the free tickets, I got to know all the carnies up and down the midway, including rides, freak shows, and "hoochie coochies." I was nearly a family name among the show people. After a week of frequenting the shows, I walked down midway early on Saturday morning before the shows opened and encountered the Ferris wheel operator. He asked me if I wanted to see something special, and being a young, adventuresome buck of seventeen, I went for it. He was oiling and adjusting the Ferris wheel, getting ready for the final big Saturday night crowd as I mounted one of the seats. He rotated my seat to the very top and stopped it. To my right on the ground some forty feet below were the show girls bathing in what looked like a water trough. They were stripped to the waist and doing their bathing with dispatch. I nearly fell out of the gondola! He must have kept me up there for thirty minutes as I gawked and had erotic thoughts as any seventeen-year-old boy might do. The ride was over all too soon, and I was back on the ground.

That day was also the grand finale of the cattle show. We had our winners and losers and those who were among the grand champion status were invited to parade the racetrack in front of the grand stand. Joe Hart took "Number Eleven," and I led Donhead. In front of me was another kid leading a yearling Guernsey bull that was as mean and nervous as they come. The kid kept telling the crowd to move back and give his bull more room. While he didn't have full horns, the calf did have spikes and could have caused harm. Following him, I had Donhead's rope over my shoulder in full confidence that he would follow me wherever I went. I didn't say a word, but when the crowd saw this black monster they folded back; I could have driven a truck through them sideways. They never knew the difference between a small Guernsey bull and a 2,300 pound Angus! Donhead was a kitten and just as docile.

Winner's Circle

Another of my great experiences was at the 1939 World Poultry Congress in Cleveland, Ohio. Three of us were selected to represent North Carolina at the Poultry Congress under the direction of County Agent, Baker Pace. We spent long hours training and learning virtually every breed of chicken in the world. I was 15 that summer and a raw recruit from the farm. My teammates were Roy Walker and Alfred Green. Both were older than I, but neither had more confidence of winning. After all, I had won virtually all my other contests. At any rate, we took off in Pace's 1937 Plymouth, up the "Skyline Drive" to Washington, then to the grape country of New York and back west to Cleveland. We spent about three days judging and sightseeing, and even though we didn't win the grand prize, we nonetheless brought back gold medals proving we were among those receiving "honorable mention." I still have mine.

In another contest, I won 200 Jersey White Giant chickens that had been sexed for all male. When they arrived as three-day olds I housed them in the brooder house in the chicken lot and let them grow for a few weeks before caponizing them. Only a half-dozen or so died as a result of the operation, and I lost another half-dozen to other natural causes. The remaining 185-190, I raised to full market size and sold half of them at Thanksgiving and the other half at Christmas. I cleared a couple of hundred dollars on that project after paying the farm for the cost of feed. I was so protective of my project that I had to shoot a stray cat that took up residence at the lot and decided to eat my profit. He didn't succeed. One premise of farming is that anything non-productive or destructive should be destroyed for the good of the farm. That held true even for domestic pets that decided to go wild.

A New Approach

In early 1940, Alton Sutton approached me and asked me to introduce a strain of hybrid corn to North Carolina. Jarvis Golden Prolific seed corn, it was called, and none of us had ever heard of it. It had been developed in DeKalb, Illinois and was quite successful there as a high-yield crop. By this time, I was working full time during the summers for the farm and had given up any thoughts of sharecropping tobacco. Sutton's proposition sounded too good to resist, so I selected the three-acre patch where V.C. and I had raised tobacco the previous year, planted the corn and tended it. "Tending" is not an accurate description of what I did. I planted it and then hired someone for a dollar a day (I made two dollars for working for the farm) to cultivate, side dress, and lay it by. I also paid to have it "pulled," shucked and shelled for the seed market. With all my labor expense, the farm's half and any other expenses, I made much more than the previous year raising tobacco and still made wages working for the farm.

4-H Club Champion

As a result of my 4-H activities over several years, I was elected County President my senior year in high school and was on the state council. My greatest personal satisfaction came in the fall of 1941 after I had entered North Carolina State. Late that summer, Sutton had urged me to gather all my records and submit them in competition for a Thomas E. Wilson Award in Beef Production. As a result, I was awarded the State 4-H Livestock Champion for 1941. This was the apex of my club work and a very rewarding way to graduate into bigger and better things. I have always been proud of my 4-H record and have Dad and Alton Sutton to thank for their long, tireless hours of training and encouragement. I still have the 21-jewel Waltham watch awarded by Thomas E. Wilson and a number of other medals from Wilson for livestock participation and winnings.

A True Prize

Andrene of Robeson (Salimony) was a Berkshire gilt pig that I won in the Sears-Roebuck pig essay contest in 1940. I got her when she was about eight weeks old and brought her home sitting on the seat of our little Chevrolet pickup, just like a person. I was always a sucker for a pretty face, and Salimony was no exception. I built a pen for her of sawmill strips about 40 feet square, complete with hog-house and trough, and fed her green corn, milk weed, slops, and all the clabbered milk she could consume. She was courted by the Beau Brummel of the county farm and, in short order, she became a grown sow and the mother of seven fine pigs.

Although I took Salimony to the county fair, she didn't place in the show, so I decided the judges were blind and entered her in the State Fair in 1941. I was so busy grooming and showing cattle that I missed her most important victory of all. My shoat, Salimony, won State Grand Champion Gilt, and I wasn't even present to enjoy my success. After showing my cattle, I hurried back to the hog barn where she stood with a blue ribbon tacked to her pen. She was the champion gilt of the North Carolina State Fair!

That same fall, I won a Holstein heifer in a judging contest. I raised her to a nearly grown cow and sold her for forty dollars. I competed in the majority of the 4-H programs: showing and judging beef and dairy cattle, hogs, and chickens. Most of the money accumulated to pay for my first year's tuition at N.C. State came either from prize money, winning contests, or showing cattle for the farm. If I was showing farm cattle, I got half the prize money. If I showed my own beeves, I retained all the prize money and premiums on sales - less the cost of purchase and feeding the stock. Anything I won through the contests was mine. All total, I accumulated about $750 by the time I left the farm to enter N.C. State.

MOVING ON UP

After my years at Bahama, school at Braggtown was paradise. I entered Braggtown in the fifth grade and blended into the class without difficulty. The children were much less competitive than those at Bahama and generally were better off financially. One thing I knew in the beginning, I couldn't lose my first fight. I didn't. I don't remember the particulars of the fracas, but I do remember that I acquitted myself well and thereby was not the "new kid on the block" to be bullied. Grade school holds no outstanding memories other than falling in and out of love several times—the girl never knowing she was my chosen one. My grades back then were average and outside interests were limited. The latter was due to the fact that we lived so far from school that I was unable to participate in sports. Moreover, I had my evening chores to perform when I reached home around four o'clock in the afternoon.

Heading into High School

Upon reaching high school in the eighth grade, I took a more active part in school life. I was on the debating team, glee club, and local 4-H club. Through the years, I remained active in the glee club, 4-H, and dramatics club. My senior year, I took a stab at my one and only role as an actor in the senior play. In my senior year, I also drove a school bus, and for that I was paid eleven dollars a month, a dollar for each route mile. Student drivers were the order of the day back in the '30's and 40's, and North Carolina drivers accrued one of the best driving safety records of any state in the Union. Many of us were experienced far beyond our tender years, for we had driven farm trucks and tractors for years. Then, too, the traffic situation was far different from today. Thirty or thirty-five student riders were on the bus, which took 45 minutes to an hour twice a day to pick up and disperse. I enjoyed my charges, although at times I had troubles with some of the "little ruffians" and had to exercise strict discipline.

Holding My Own

In order to establish authority, I had to occasionally take measures that today would be unthinkable. Once, a couple of toughies threw an apple core at me, hitting the back of my head while the bus was in motion. I stopped the vehicle and promptly threw the little devils off at least a mile from their home. The extraction of the two of them was no small order at best. The next morning, as I approached their pick-up point, I spied their "gorilla" brother standing with them, obviously to give me a work-over. I didn't even slow down. Upon arrival at school, the P.A. system summoned me to the principal's office where mama, papa, and "gorilla" sat. The principal wanted to know why I had thrown the kids off and refused to stop that morning. I carefully explained the problem and told him that if he expected me to operate the bus in a safe manner, he'd have to allow me to act in the best interest of my passengers' safety. That was the last I heard of the incident, and the two children behaved beautifully the remainder of the school year.

In a Great Groove

Unlike many of my classmates, I had a deep love for school. Not that I was all that academically inclined, but when September rolled around and school started, my workload on the farm diminished accordingly. Also, living way out in the country meant that my social life as a young boy was zero. School provided me the opportunity to mingle with my classmates and provided a sense of belonging. Although I didn't start at Braggtown until the fifth grade, I spent the remainder of my school years there until graduation in 1941. Many of those I met in the fifth grade were still classmates at graduation: Terry Beasley, Sally Vaughn, Hilda Rakes, Kathleen Woody, Wayne Tilley, Clifton Garrett, and others whom I cannot remember. We received an influx of grammar school graduates from other school when we arrived in the eighth grade: Anna Waddell, Edwin and Melbourne Nash, Bobby Miller, "Tookie" Browning, T.B. Wheeler, and others. There were 51 seniors in my graduating class, the last class to graduate during peacetime for several years.

High School Graduation

My most memorable part of graduation was having Dad address the graduating class. I don't remember a great deal of what he said, but was so proud of him for giving an excellent address with a straightforward delivery. At the class' request, I asked him if he would be the commencement speaker two months in advance, to which he readily agreed. At least once a week, I asked him if he was writing his speech. His reply was always, "I'll get around to it." By the last week of school, no speech had appeared, and I was getting panicky. At the exercises, he got up and made a beautiful delivery. You see, I had forgotten that he had been a schoolmaster in his earlier days and was quite comfortable before a student audience. No doubt he had given several variations of that talk years before.

College Cut Short

While I never claimed to be an apt student, I do feel that I was woefully unprepared for college. Certainly, I would not deny that my inattention was partly responsible, but mostly, I feel that it was due to the lack of adequate instructors. Some of our teachers were dedicated, but many of them were just putting in time and drawing a salary, which was meager at best. Adding to the teacher problem was the fact that we attended an eleven grade, eight month a year school. I had never had Chemistry, Solid Geometry, or other basic courses in preparation for college. In the fall of 1941, upon entering N.C. State, I was a lost ball in high weeds! To make matters worse, I had elected to study Aeronautical Engineering. Had I stuck with Agriculture, something I knew, I am sure I could have done much better, but I had always wanted to fly and had a great love for airplanes. Well, it didn't really matter anyway, because World War II broke out that fall, and I was saved by the draft a year later.

REMINISCING

I spoke earlier of the beauty of Fairntosh. Let me reminisce. The summers were hot, and the winters were cold. We didn't have central heat or air conditioning, except the breeze that God provided. I occasionally saw snow on the windowsill and floor inside my bedroom. I have also experienced insufferable heat so stifling that it was impossible to sleep. On those occasions, I often took my pillow and slept on the downstairs floor of the front hall with the front door wide open. I also recall the many times that I would go outside under a full moon and seek the beauty of the "Big House" lawn, complete with whippoorwills or hoot owls, or whatever night noises prevailed including the lowing of cattle, dogs barking, or a gentle breeze blowing through those magnificent trees. Our property had white oak trees that were a couple of hundred years in age. Many of them have disappeared in the past half-century, but not in my memory. I remember climbing into their uppermost limbs and hiding when Mother would call: "Glenn, are you in that tree again?"

Time Flies...and So Does the Snow

I often imagined myself in a small homemade aircraft, taking off in a narrow space in the chicken lot, soaring over the lawn, down over the barnyard, barns, and down into the meadow to a landing on the old horse racetrack. I guess I have flown that pattern a thousand times. Likewise, when I tired of the drudgery of the tobacco field, I could rest at the end of a row on my back and look up at the billowing thunderheads, imagining myself in a plane, soaring in and around those beautiful clouds. I had no idea of their treachery at the time!

In the early winter of 1940, we had a blizzard that lasted a week. Snow began to fall in early afternoon, and school was dismissed so we could get home before trouble really started. By morning, we had ten inches on the ground, and drifts of three to four feet. Of course, the only work that day was to tend the livestock. We sat around the fire and enjoyed the cozy feeling of warmth and plenty while the cold prevailed outside. The temperature dropped to minus eight degrees or so for several days. (That was the time Gus Haskins and "Nubbs" Hackney engaged in their altercation.) Late the first morning, Dad suggested that I crank that tractor, attach the road scrape, and scrape the snow off the lane out to the main highway, assuming that the county would have cleared the road by then. Well, I went down to the tool shed where the tractor was housed and cranked it. When I climbed aboard and put it in gear, it simply would not budge! No matter how much throttle I advanced, it didn't move. Then I realized that the water in the tires (used for traction) was frozen. After several attempts of rocking it back and forth, I finally got it to move a few feet, until the ice was on top of the tires. At that point, the tractor jumped about six feet—liked a scared kangaroo. Fortunately, the doors of the shed were open. I had to reverse the process, build a fire and allow the heat of the stove in the shed to melt the ice in the tires before I could proceed to scrape the lane. It was late afternoon before I accomplished a job that should have been finished by noon.

The snow did provide recreation, too. I learned another trick of hunting without a gun. By taking a tobacco stick and walking across the snow in the pastures and sage fields, I could pick up rabbit tracks and follow them to the end where a tired and hopeless rabbit awaited his fate. We ate rabbits, squirrels, and all game—except possums—at that time. No one had ever heard of rabbit fever then.

One of the days during the cold snap, Dad had gone to Durham in his car, leaving the pickup in the barn. By the fourth or fifth day, the snow had thawed and melted until it was a solid sheet of ice. We boys got the pickup out in the pasture where there were no obstructions, and gunned it up to twenty or thirty miles per hour, spun the steering wheel, and slammed on the brakes. That pickup would do about three complete loops. Talk about a wild ride, wow!

One is the Loneliest Number

In my early teens, I must have been the loneliest kid in Durham County. At thirteen or so, I knew what girls were, but didn't have the age or experience to do anything about it. I spent hours on weekends tearing down and rebuilding a small half-horsepower engine that had been used on the well pump, just to kill time and see if I could do it. In the summer, when hunting was not in season, our German shepherd, Fritz, and I would roam the woods for adventure. Finding boats on the river that had broken loose during the floods above the dam and survived the spillway, finding adventures that only a small lad can imagine, and generally exploring the entire six thousand acres kept me sane.

While horseback riding was my penchant, I had not yet graduated to Nancy. Bessy was a good horse but wasn't gaited and had little spirit. So, in order to add zip to the ride, I built a small wooden sleigh, just large enough to sit on, and tied it to a plow line around her neck. Then I attached the lines to her bridle and drove her all over the farm, riding on my sleigh. That was in the summer, and the wooden runners didn't last long on the gravel roads. That didn't matter. When that one wore out, I simply built another.

SOME HISTORY

While most of the information set forth here is of my own recollection and research, I also give credit to Charles Richard Sanders, author of "The Cameron Plantation in Central North Carolina" (1776-1973) for dates and much of the chronology. I lived at Fairntosh Farms during my youth, from 1931 through 1941, and recall much of its history as told by many of the elder citizens, some of whom had been slaves prior to the "War Between the States." It is a good story, filled with successes and failures, hardships and slavery, riches and tragedies, and of the five families who ruled and owned the largest plantation in North Carolina.

Fairntosh Heritage

Duncan Cameron, in his mid-20's, riding his horse from his parental home in Lunenburg/Mecklenburg Counties, Virginia, crossed the Flat River in North Carolina, which separates Granville County from Durham County, on his way to Hillsboro, where he practiced law. When he reached the Eno River to the North, he found it in flood stage and was unable to cross. Retracing his way along the "Old Indian Trail" back to the Flat River, he found that the rains had also swollen it to the point of being impassable. All this while, he had been traveling on the plantation of a wealthy farmer/merchant by the name of Richard Bennehan, who lived in his modest but comfortable home, "Stagville," with his family. Calling on the Bennehans, who gave him lodging for the evening, he discovered that they had a lovely daughter by the name of Rebecca. Wedding bells chimed, and they were married on February 24, 1803.

Richard Bennehan (1743-1821) was the son of Dudley and Rachel Bennehan of Richmond County, Virginia. He moved from Petersburg to what is now Durham County, and purchased land on the Flat River and built his home, which he named Stagville. He was also a successful merchant, having stores in several locations in partnerships with others. In her dowry, Rebecca received as a gift from her father three hundred acres of the plantation where the newly-weds built their imposing home and named it Fairntosh, after Duncan's father's ancestral home in Lochiel, Scotland—Feirntosh.

Duncan Cameron (1777-1853) moved to Hillsboro in 1797 at the age of twenty to establish his law practice. He became quite successful as a businessman, statesman, lawyer, farmer, and banker. He and his new wife, Rebecca, built Fairntosh mansion in 1804, the year following their marriage. He was Chairman of the Building Committee that built the N.C. State Capitol Building. Always involved in education, Duncan served as trustee of The University of North Carolina for 36 years-from 1802 to 1838. He served as Judge of Superior Court and a Director in the State Bank of North Carolina. The union of Duncan and Rebecca produced eight children. However, as the only surviving son, Paul Carrington Cameron inherited both Stagville and Fairntosh.

Paul Carrington Cameron (1808-1891) had seven siblings: a brother who died in infancy and six sisters. Being the surviving male heir, as was the custom, he inherited the bulk of his father's estate. Being industrious and an entrepreneur, he built the land holdings into a massive kingdom of 60 square miles of farm land and forest, embracing some 30,000 acres in Durham, Granville, and Wake Counties. He also had 1,900 slaves to till the plantation and serve the needs of the family. Paul Carrington married Anne Ruffin in 1832, and their union produced ten children, seven of whom survived to adulthood. Paul Carrington was, without question, the single wealthiest individual in North Carolina and one of the most influential. Like his father, he was a Trustee of the University of North Carolina for 27 years and was still on the Board at his death in 1891. He also served on the Boards of several railroads serving communities in North Carolina. His great interest, however, was in the development and nurturing of his Fairntosh property: including a woolen mill, cotton mill, granaries, and cattle. Much of his estate was lost to "Yankee Carpetbaggers" and squatters following the War Between the States. The plantation still embraced over 6,000 acres when I lived there in the '30s.

Bennehan Cameron (1854-1925) was born in the mansion at Fairntosh. In the division of his Father's estate, Bennehan inherited Fairntosh (which included Stagville) and what was left of his father's holdings following the Civil War. He married Sally Taliaferro Mayo on October 28, 1891. Their union produced two daughters: Isabelle and Sally Taliaferro. Bennehan, like his father before him, was a railroad developer and was instrumental in building the North Carolina Railroad, which later was (and still is) leased to the Southern Railway (now Norfolk Southern.) He was a banker and a Trustee of the University of North Carolina from his father's death in 1891 until his own demise in 1925 - three generations of service to the cause of higher education. Bennehan was a farmer, although perhaps not as dedicated to the profession as his father. His love was horses, especially race horses, to the extent that he built stables and a racetrack in the lower pasture (still visible when I lived there) and a cupola on the roof of the "big barn" from which he could watch his horses work out. Upon his death, the plantation began to fall into disrepair.

Stagville Heritage

Isabelle, known to us as "Miss Bell," married late in life and, to my knowledge, left no offspring. She sold her half of the plantation, Stagville, to Liggett & Myers Tobacco Co. after the daughters divided the property. Sally Taliaferro, known to us as "Miss Sally," married John W. Labouisse, a successful cotton broker from New Orleans and Charlotte, and upon his retirement in the late 1950's, they remodeled the mansion with modern conveniences such as lights, water, and heat, and lived there until he died. Sally then placed the Fairntosh portion on the market. It was purchased by a nearby tobacco farmer by the name of McFarland for a "song." Following a few years of hardship, he sold his portion to former Governor/Senator Terry Sanford and his group of investors, who developed it into a "total community" named Treyburn. At the time the Sanford Group acquired the McFarland Property, they also bought the land owned by Liggett & Myers, embracing a total acreage of over 5,000. The "big house" was retained by the Consortium, and the Stagville home was donated to the State as a Historical Preservation Center, along with the barn and slave quarters at Horton Grove.

Fairntosh has had a long and interesting history. Although I have shared my part of the history growing up in the 30s and 40s, following are some additional memories worth mentioning.

BITS AND PIECES

The Canoe

In the summer of 1940, I believe, I dreamed one night of finding a canoe on Flat River. The next morning, before breakfast, I told Mother of my dream. I described just how I had come upon the canoe in my dream and found it lodged on a pile of rocks in mid-stream with a hole caved in on one side, and filled with an inch or two of water. In my dream, I then went to the bank of the river and found a double paddle resting on a pile of debris. Mother responded with an old wives tale, “Friday night dream, Saturday morning told, will come to pass before it is nine days old.”

I had already found several wooden boats, all of which had washed over the dam upriver, and I had one tied on the river at that time. After breakfast that Saturday morning, Dad, V.C., and I decided to go fishing, something I cannot remember our having ever done before or after. When we rounded a bend in the river, there on a pile of rocks was the canoe from my dream! It had a couple of broken ribs and a six-inch gash in the canvas cover, and was filled with a couple of inches of water. My search for the paddle, however, proved futile—keeping my dream from becoming a vision! It’s an odd story, but it did happen just that way.

Forest Fires

Numerous forest fires burned over Fairntosh during the ten years we lived there; some started by lightening, but most caused by careless hunters. On a 6,000-acre farm with perhaps 5,000 of it in grown timber, a fire could be devastating. It took tractors, wagonloads of water, and all the men we could summon with backpack water sprays to put out the fire. Sometimes, they would rage for a day or two before they could be brought under control. They were never as dangerous as those out west, simply because our woods were not as dry, and the fires burned only the underbrush and never "topped out" as most of the western fires did. That is, the flames never reached the top of the trees and jumped from top to top. They were, however, awesome and dangerous to man and animal. While the land was posted, we had numerous poachers who came from the surrounding neighborhood, especially Durham, to hunt because of the abundance of game. We had an old iron bell on a pole at the big barn, which was only rung when fire was discovered to summon help for fighting the forest fires.

Riding the Trees

I remember all the rides at the State Fair, but none were more exhilarating than the ride on a tall sapling. The exercise was to get a group of neighborhood boys together on a Sunday afternoon, looking for an adventure and mischief. Four or five boys would do. We would find a tall hickory sapling out by itself and all climb it together. By sheer weight, we would bend it to the ground, where all would get off—except one! A wilder ride there never was, especially when the single rider reached the end of the first "whip back." If he didn't stick, he'd be catapulted into tomorrow. Boy was it fun!

Building Sand Forts

While firecrackers were relatively inexpensive during the depression years, few people had money to buy them. We made the most of what we had, both in resources and imagination. We would build sand forts on either bank of a creek, elaborate with moats, drawbridges, flagstaff, and blockhouses. Once constructed, we'd place the small two-inch firecrackers under the fort walls on both forts and see who could do the most damage. The game was to blow up each other's fort. The winner was the builder of the fort sustaining the least damage.

Oscar

When I was perhaps twelve or thirteen, a friend of mine and I went into the woods on an Easter Sunday afternoon looking for young squirrels to tame and keep as pets. We found a couple of small squirrels about three weeks old in a nest. They were too small to be of much danger and not old enough to have become accustomed to the wilds of the forest. He took one, and I the other to our respective homes to raise. I don't know whether or not my friend's squirrel survived, but mine did. I named him Oscar, after a skeleton by the same name in our science lab at school.

As a tiny kit, I had to feed Oscar with a medicine dropper filled with warm milk, preferably right out of the cow, until he got large enough to eat peanut butter and progress to more solid foods. As he grew, he became a household attraction, running and jumping from furniture to the tops of doors and onto an unsuspecting back for a free ride across the room, only to jump off at the next high point. He literally had the full run of the back part of the house, and was greatly enjoyed by the whole family.

When he was just a kit, I made him a small wire cage that we kept in the kitchen with enough cloth for him to nest down in at night. On cool spring mornings, I would take him out of the cage, wrap him around my neck under my jacket collar and take him to the barn with me while I milked the cows. He would sleep the entire time. As he grew and ate a greater variety of foods, we fed him most anything we ate; cookies, corn bread, seeds of all kinds, and most of all, peanut butter. About the only thing we didn't feed him was vegetables. He'd sit on his hind legs and eat cookies by twirling them in his forepaws as he nibbled.

By early fall, Oscar had grown to almost full size and in doing so, grew more aggressive. He had sharp teeth—his favorite weapon. I realized that it was time to build a larger cage and keep him outdoors. I constructed a cage approximately four feet cubed, complete with hollow stump and a roof, all on legs two feet off the ground. He escaped once for a couple of days, but when he got hungry, he came up to Mother, who was working in her flowers, and begged for a handout. She promptly fed him and put him back into the cage.

By early fall, he had bitten two or three people, including me, and must have picked up an enemy somewhere along his defensive road. One morning I went out to feed and water him, and found Oscar curled up dead, obviously at the hand of someone who had poisoned him. I don't look at the morality of whether we should have taken him from the nest in the first place, but I can say that he gave our family many hours of pleasure and a belly full of laughs.

Fritz

Then there was Fritz, our German shepherd dog who was the master of everything around Fairntosh. Fritz slept under the door of the chicken house or on the front porch and watched everything about the farm, day and night. As a matter of fact, we never had a lock on the hen-house door, didn't have to. On hot summer nights when he slept on the front porch, he would raise his head at the slightest noise and listen. If he was content with what he'd heard, he'd go back to sleep. If not, he'd go and investigate. If he barked, it was a good idea to check it out because he had found something he didn't like.

I've already told of the 200 capons I raised in a 4-H program. We kept them in the chicken lot but, on occasion, one of us would forget to latch the gate and some would escape. One Thanksgiving afternoon, after the family was stuffed on a wild turkey Dad had shot, I looked out in the side yard and saw three of my capons lying out there dead. I went outside, gathered them, and laid them on the back porch, wondering what on earth caused their demise. A short while later, I found three more dead capons. I was about the hit the panic button when I heard a commotion on the other side of the house. My investigation revealed Fritz carrying one of my capons by the neck back to the chick lot gate. When they stopped struggling and died, he would drop that one and go after the next. They were not supposed to be outside the lot, and he was there to see to it that they weren't. To this day, I don't remember what we did with seven dead capons.

Fritz followed the wagons to the low ground daily in the summertime and put in as many hours in the fields as did the other help. The temperature would get insufferably hot down on the riverbanks in July and August. He would hunt a mud wallow in the road, filled with a few inches of water, and stretch out full length in that pool. When he got up, he would almost tiptoe out of the wallow to keep the water from draining off his fur. Most dogs will shake themselves when they get wet, but not Fritz. He knew the water would keep him cool longer if he left it trapped in his coat.

He and I would range over the entire farm when I was small, exploring, investigating, just enjoying nature. He was as docile as a kitten to those of the family, but he could be as vicious as a wolf to strangers. I saw him tackle Gus Haskin's pack of hounds one morning and whip them all, including Gus! If a stranger came to the house, Fritz would not let him out of his car until one of us came to the door and spoke to him. To say that Fairntosh was well guarded would be an understatement.

When we left the farm, we took Fritz with us to Durham, but had to chain him in the back yard to comply with Durham leash laws. By this time, he was old, wrought with rheumatism and fleas. We finally held a family meeting and decided that the most humane thing to do would be to put him out of his misery. I think that was one of the saddest days of my life!

Hog Killin' Time

On cold, brisk fall mornings with the temperature in the twenties, we would get out the scalding vats, build a fire under them, sharpen our butcher knives, and prepare for a "hog killing." The hapless beasts were slaughtered by cutting the throat and running a sharp knife into the heart. They were allowed to get up and literally walk themselves to death. Then came the scalding and scraping, hanging them on gamboling polls and swinging them by the hind legs to the "A" frame for gutting. The dressed hogs were then loaded on the bed of the pickup and transported to the meat house where they were dissected, trimmed, and ground into sausage or the meat salted out for curing. Some of the hams, shoulders and side meat were taken to the smokehouse for curing while the rest was processed at the meat house. It was a nasty, back-breaking job that lasted from sunup until late into the night.

The side meat, jowls, tails, and feet were salted down in the bin to be sold to the tenants the following summer when their meat gave out. Hams and shoulders received further treatment with brown sugar, sage, pepper, and salt before wrapping and hanging them in starched white flour sacks for curing. They would age a year before being sold. We sold meat, kerosene, and molasses for ten cents a measure—be it a pound, gallon, or pint.

In our earlier years, we peddled the fresh meat in Durham in the black section of town by going from door to door. Fritz always rode in the cab of the "A" Model Ford pickup during these excursions. It was a good way to raise ready cash during the depression years and helped pay the farm bills.

Dr. Taylor

As a boy, I took it upon myself to doctor every sick animal that the farm produced: chickens, cows, pigs - the whole menagerie. That's when I thought I wanted to be a vet. By the time I was fifteen or sixteen, I realized the filth and gore that went along with doctoring sick animals. I had done it all, including midwifery to a ewe that was having twins simultaneously. I set up shop at the house and doctored "runt" pigs, sick chickens, and the lot. One of my patients was a pullet that couldn't digest its food out of the craw. Weekly I operated and removed excess grain and grit from the craw only to have it die of starvation. I named it "Bunchy" because of the excessively large craw. After Oscar the squirrel died, I turned his cage into a regular zoo, capturing owls, possums, crows, and other wild creatures. Eventually, I let them all go.

Payday

At about the age of 16, I had decided that raising tobacco was not for me. I so informed Dad, who asked what I was going to do during the summer, to which I replied, "Work for the farm." As straw boss, I made two dollars a day. On my first payday, I stepped up to the pay desk as any other hired hand, and Dad handed me a five and two one dollar bills. Something was wrong here; I was making twelve dollars a week and only got seven. Dad patiently explained that since I worked for the farm, I had to pay room and board just like all the other hands that had worked and lived at our house. He pointed out that Mother had her expenses for food, paying Aunt Judy to wash and clean, and now that I was older and working for the farm, surely I must realize that I could not be treated differently. I understood that, didn't I? Well, I didn't! I thought to myself, "Now, there is the meanest man that ever lived, charging his own son room and board!"

By the second summer of this practice, I had become accustomed to the CUSTOM and said nothing. When I entered college in the fall of 1941, Dad handed me a saving account book that registered five dollars a week for two summers' work. His remark: "I knew you wouldn't save it, so I did it for you."

Country Cookin'

Mother cooked on a wood range. It was made of cast iron, and sat in front of the fireplace, which had been covered with sheet-iron. The stove flue connected the range to the fireplace for smoke outlet. The range itself was an engineering masterpiece. Fueled with sticks of pine stove wood for quick heat, it heated not only the stove surface, but also the oven, hot water box, and the warming oven which was located above the stove. My favorite place as a small boy was to stand in the wood-box behind the stove on a cold winter morning and watch mother perform her magic. Ah, the food that came off and out of that stove!

First, was the country breakfast: ALWAYS hot biscuits, milk gravy, and eggs. Meats consisted of ham, tenderloin, sausage, thick bacon, fried chicken, or pork chops. Often, the above was topped off with a large portion of fried apples. To conclude the breakfast required another biscuit smeared with country butter and an assortment of jams, jellies, or preserves. We enjoyed that type of meal EVERY morning!

Lunches were somewhat varied, due to the season. Of course I attended school eight months out of the year and only ate lunch at home on weekends and during summer "holidays." The lunches I can remember most were those prepared by mother in the summertime. Consisting of fresh vegetables, meats, pickles, or tomatoes, topped off with a glass of cold whole milk. Cornbread was a lunch staple! The vegetables could consist of an assortment of either snap beans, field peas, cabbage, turnip greens, collard greens, butterbeans, Irish or sweet potatoes. I am sure we ate others, but I have forgotten – Mother cooked so many different dishes. Meats were not as prevalent during the noon meal; perhaps some fatback cooked into the vegetables, or an occasional side meat, but mostly, that was saved for supper.

Having had two heavy meals already, often supper was light. Dad suffered most of his adult life with ulcers (maybe it was all that rich food) so Mother cooked the evening meals mostly to pamper his stomach. Homemade cottage cheese was a regular, as was cornbread and potato soup. Sometimes Mother would bake fresh rolls, and we would feast on butter, rolls, and assorted jellies or molasses. This fare changed somewhat in the wintertime, since my brother attended school and didn't get to partake of the usual noonday meal. That being the case, we often had another lunch-type meal at suppertime. As a little fellow, I was allowed only milk at meals; however, as I reached my teen years, I drank either ice tea or coffee.

Then, there were the "game" meals that we harvested off the farm: quail, rabbits, squirrels, and turkeys. We often had fried rabbit, squirrel, or quail for breakfast, and usually had a wild turkey for Thanksgiving and Christmas – with all the trimmings. On those festive occasions, we were allowed a small glass of Mother's blackberry or scuppernong wine.

And all this we ate during the depression years!

Washin'

Everybody knows that Monday is washin' day – just like Saturday night is bath time. But, few of the modern generation know just what washin' entailed sixty or seventy years ago. Today, with the washing machine/dryer combinations, washing dirty clothes is a simple matter of sorting and placing the dirty clothes in a washer for a few minutes and -- voila -- you have clean laundry. Not so when I grew up.

First task is not washin', but "hog killin'." Yep, you didn't buy laundry soap, you made it. In the process of slaughtering and working (cutting up) the pig, a decision was made whether to render the fat into lard or lye soap. Now, lye soap is much better than say Octagon, for the simple reason that it is stronger and gets more dirt out of farm-soiled overalls, shirts, and socks that were worn most of a week without change. Lye is added while the fat or lard is in the cooking pot and allowed to cook and cool until hardened. Then, it is cut into chunks and sorted for later use.

Now let's get to the really hard part. On Monday morning, a fire is set under the iron wash pot, and when it is boiling, the dirty clothes and lye soap are added for boiling. Meanwhile, there are two number three wash tubs standing on either side of a wringer frame, ready to receive the boiled clothes when deemed ready. That in itself takes skill. The washer doesn't have to sample the clothes; he or she simply knows when it is done.

The clothes are lifted out of the wash pot with a wooden spatula (which was used to periodically stir the clothes while boiling) and placed in one of the awaiting tubs, complete with washboard, ready for further scrubbing. I often wondered how the "wash woman" could tolerate her knuckles scrubbing on the washboard, but that's what it took to get those clothes clean. Once finished with the scrubbing, the clothes were then wrung through a hand-cranked wringer into the other washtub filled with rinse water. Following that, the process reversed, and the clothes were wrung back through to the first tub, which had been filled with bluein' water. Now, I never exactly knew just what bluein' water was for, but Mama said it was to get the shirts and underwear white. I don't exactly understand how bluein' water turns things white, but she said it did, and I didn't argue with Mama.

Once through with the second rinse (bluein') and once again wrung, it was time to hang the laundry on the line to dry. That worked well on a hot summer day when the sun was shining, so it always became necessary to wash on a clear Monday. Then in the wintertime when the temperature was below freezing, once the clothes were hung, they freeze dried! I didn't understand that either – except, when they were brought into the house, they were stiff as a board, and upon thawing, were dry. A miracle I'm sure.

Most everything was cotton that was washable and had to be ironed. In the early years, we didn't have electricity, and the iron was a "flat iron" heated on the wood-burning kitchen stove. The clean laundry had to be sprinkled before ironing, and the shirts were treated with starch for the collars, folded, and placed in the dresser drawer for use next week.

Sounds primitive, doesn't it? Well, it was! But, that's how laundry was done in the "good old days."

Surrey County, North Carolina

During the summer of 1929 just prior to "Black Tuesday," and two years before living at Fairntosh, my family moved to Surry County from Lattimore, NC. Dad became County Farm Agent following several years as Vo-Ag teacher in Cleveland County. I was five years old and had not yet started school, as there was no pre-school or kindergarten in those years. While I doubt that I was old enough to recognize the degree of poverty in Surry County, certainly my family was not rich either, and I did begin to notice the difference between our new home and Cleveland County. Tobacco was the cash crop in the former, where cotton was king in the latter.

Then came "The Great Depression" in October, and the bottom fell! Poverty ruled Surry County with the price of tobacco falling to unprecedented lows. Houses that needed painting weren't. Roads that needed improvements were left unattended. Schools and churches suffered deferred maintenance. Other than the fruit, vegetables and meats grown on the farm, people had little to eat. My family was more fortunate; Dad had a salaried job that paid fairly well, considering the times, but lived in a rented house and drove a 1927 Essex "Super-six." Our house was owned by a Miss Cooper, who taught school in Eastern Carolina during the school months and occupied the back upstairs apartment only during the summer season. Her brother occupied a single room; the rest was our living quarters – without central heat or plumbing. We did have electricity, but the outhouse was in back. We had access to a barn where we kept our milk cow Flossie. All in all, we had it much better than most.

Some of our relatives lived in the county and were tobacco farmers. Their lot was about equal to the majority of farmers in Surry, who were struggling to keep their heads above water. The roads became almost impassable during winter months; consisting of "hog-back" red clay (few paved roads existed) and hills. Many roads were without bridges and streams were forded by both automobiles and wagon teams. Homes and outbuildings fell into disrepair, as there was no money to buy lumber or paint to maintain them.

Later, in 1931, we moved to Fairntosh in Durham County, and saw similar conditions there; though not as severe. Automobiles gave way to "Hoover carts" (the rear axle and springs of a Model "T" with rubber tires and a makeshift seat) as a means of conveyance. People could not afford gasoline at fifteen cents per gallon or the cost of repairs of their automobiles, so they hooked a mule to a "Hoover cart" and drove to their destination. It was commonplace to see a family ride to church in a wagon, in which straight-back chairs were added for women. The little children rode on the back with their feet dangling.

Then came World War II. Almost immediately, everyone who wanted a job had one. Factories were booming with war production, making guns, ammunitions, airplanes, tanks, and ships. Many went off to war as soldiers, sailors, or Marines, but others stayed home to man the factories that supplied their loved ones. Of those who did stay home, some remained on the farm and worked two jobs – the defense plant and farm. Money then became more plentiful. With the advent of wartime rationing, those on the home front didn't have a great deal to spend their monies on, so they saved their newfound wealth and bought War Bonds.

Then, hostilities ceased in August of '45 with the advent of the "A-Bomb" and the troops came marching home to get married, raise families, and buy homes and cars. Those in Surry County went back home to a better life. They rebuilt their homes, or as is evident now, built new ones near the site of the old one. Roads were surfaced with Kerr Scott's "farm to market" roads. It became easier to get out of the hills and go to the market to sell their tobacco and buy the goods necessary for a better life. The G.I. Bill educated many in new farming practices, on the job training, and college.

As I travel Surry County today on my annual pilgrimage to the Taylor reunions at Union Primitive Baptist Church, I often reflect on "how it was and how it is." Patiently, Sarah Jane listens to me tell the stories for the "umpteenth" time. One thing that strikes me as being significant: I don't see nearly as much tobacco being grown there now as in by-gone days. In its place are pastures and cattle, row-crops and chicken barns. I believe Surry County has arrived! Those rolling foothills to the Blue Ridge Mountains are beautiful.

"MAKING DO" WITHOUT ALL THE MODERN CONVENIENCES

I was a child a long, long time ago. I am now halfway through my eighty-seventh year, which means I have lived approximately thirty-six percent of our nation's history. If you don't think that is a long time, let me reminisce with you for a few minutes. Oh, I would have to start with the horse and buggy era. No, I didn't go to church or school in a buggy, but I well remember those times when others did! I had a great uncle (Grandmother Taylor's brother) who was a doctor and traveled in his buggy as late as the nineteen thirties. Model T Fords were prevalent in those days, and the "spiffiest" car on the road was a 1930 Ford Sport Roadster convertible. Mules and wagons were the trucks of that era, and no good farmer could be without a good team. They were used for plowing, both to break the soil for fall conditioning (with a two-horse plow) or spring cultivating with a one-horse plow. Tractors and mechanized farming were a rare luxury in those days.

The Great Depression

The Great Depression of the 1930s hit extremely hard, but a farmer with the above equipment on a one hundred acre farm could subsist. He grew his own food and "cash crop" to carry him through the year until the next harvest. Life wasn't easy. Many "city folks" had a rough time of it, unable to get a job and having little cash to provide their families with food and clothing, not to mention housing and utilities. Many went broke while others became hobos and rode the rails to distant places looking for a job, an escape, or something else, which they could not define. In my own family, we (Dad) purchased a Philco radio about 1928 and strung wiring from the house to a tree to provide airwave reception. This was a luxury few could afford. We listened to Amos and Andy and other programs, which I as a four-year old could hardly fathom. Our radio stations were all fifty thousand watts and scattered throughout the Eastern United States: KDKA, Cincinnati; WWL, New Orleans; and WPTF, Raleigh. We did have electricity and a well pump on the back porch and outhouse in the distant back yard. We heated one room with a fireplace or heatrola (wood stove), and mother cooked on a kerosene-heated kitchen stove. Food in our home was simple but nourishing.

Oh, and we did have a four-party telephone line where our parents could make a call to neighbors (if the line wasn't busy) by twisting a crank on the oak wall box phone. Nah – we didn't use smoke signals; that went out with the Indians! Our toys were simple and required much imagination. We used sardine cans with the top rolled forward to act as a convertible, and a large medicine bottle was used as a bus or truck. We built roads in the dirt and slid imaginary vehicles down our road crawling on our hands and knees. Mother patched a lot of knees on our overalls.

Boys Will be Boys

My brother and I were fortunate that our parents had given us a Red Flyer wagon, and we hauled firewood to the house in it, plus pulling each other around the back yard. Some of the neighbor boys were more ingenious than we were – and older. They used old worn out tire carcasses as a mode of transportation. It required two boys to play – one to get inside the tire, all wrapped around the inside rim – and the other to give him a hefty push! The rider, when proficient, could keep the tire rolling by exercising inertia with his body. I never learned to do that! Our next door neighbor, a kid named Washburn, had a tree house, and we spent much time climbing up in it as our club house. On another occasion, a neighboring boy decided to learn how to fly by jumping out of the barn loft onto a pile of hay using a plank wing. This didn't work, and he broke his leg.

As we moved on to another community, we grew older and bolder. Moving to Dobson meant having numerous cousins who became our playmates. Billy and Julian, who lived at the old Taylor home place in the late twenties, were ingenious if not innovative – or both. They built a bag swing out over a hillside, tying a plow-line to a limb high up in a tree. We would run and jump on the bag, suspended from the rope and swing out over the hillside. What a ride! I did the same thing some years later by building a scaffold to get height before jumping on the bag swing. The only problem was that I used a chain instead of a rope. The chain broke and left me treading in mid-air about fifteen or twenty feet off the ground.

But not for long!

The Taylor cousins would build sleds out of wood and scoot down the hill on pine straw, not unlike snow sledding. Or they cut two-inch slices off a pine tree, drilling holes in them for wheels and attaching them to a scooter frame. A pin in the front axle would allow the rider to guide the vehicle with his feet. This, too, was a harrowing experience, keeping the scooter from hitting a tree. We also dug caves into the clay bank along the road on "Kid Hill" just below the family farm. It never caved in (I'm not sure why), but that was the challenge of the cave. Earlier, in Lattimore, some of the older neighborhood kids dug trenches along a dirt mound that had been excavated to dig a basement for a country store. They used old tin signs to cover the trenches and covered them with clay dirt to make an awesome cave. Mother threatened us with physical harm if we tried to enter that cave. I did. Six times, as I recall, and got six spankings!

Well, that will do for starters, but I thought it would interest you to know what we did before the days of computer games, Lacrosse, football, skateboards and roller blades. See, we had fun too!

REFLECTIONS OF CHILDHOOD

In the fall of 1990, I again toured Fairntosh with Sarah Jane and my cousins, Peggy and Ben Winstead. The day was a beautiful autumn afternoon, full of color and sunshine. We approached the farm from the south, crossing the original boundary, during my time, at Little River. Upon arriving at Eno, we noted that many of the slave quarters had rotted and crumbled to ruins. A new road had been built directly from Eno Quarters to Harris, where I labored so many hot summer days in my tobacco patches. Covering the entire area of Harris now is a large industrial complex of several hundred acres. Roads have been located and buildings abound on both sides of a gigantic loop. I couldn't help thinking of the numerous Indian artifacts that must have been covered or destroyed, for this is the area where I salvaged so many Tuscarora arrowheads and spear points. I suppose in eons to come, when the industries have ceased to exist, those artifacts will still be there for future generations to discover. Still, it's a pity they are lost for now.

Fairntosh and Stagville were split by the Cameron sisters following World War II, and one of the sisters sold the Stagville portion to Liggett & Myers Tobacco Company. The other sister and her husband, John Labouisse, refurbished the Fairntosh mansion and moved up there upon his retirement in the early '50s. Upon his demise, she sold the remaining 3,000 acres to a local tobacco farmer named McFarland. He later sold it to a consortium put together by ex-Governor Terry Sanford. Sanford's group bought Fairntosh, Stagville, and a portion of Snow Hill to develop a total community north of Durham. Some years ago, L & M Company dedicated the old Stagville mansion, some acreage, and the Quarters and barn at Horton Grove to the State of North Carolina for use as an archives and study center of antebellum mansions within the state. It is widely used by both historians and architects. Sanford and his group cut out the Fairntosh mansion, complete with pastures and outbuildings, and preserved its beauty and antiquity. The balance of the property of 5,000 acres is in the process of development. In addition to the industrial complex at Harris, the consortium has built a dam on Little River and is developing a residential community complete with golf course, clubhouse, and all the amenities of affluent living.

From Harris, we drove to the "Big House" at Fairntosh and asked for permission to explore the grounds. Although the Sanfords were not there, those present were courteous and helpful. We found the houses and grounds to be in excellent repair and a continuing program is being followed to maintain the estate in top shape. Some of the outbuildings have been modernized and are used as rental units for professors and other professionals from Duke and Durham. A half-dozen of these buildings are now in use. It's a beautiful setting and who wouldn't give his right arm to live in that environment! The big barn is in need of repair, but is scheduled for renovation in the near future, as funds become available. A couple of the new barns, built since I left Fairntosh, have been renovated and now house a small herd of beef cattle.

From Fairntosh, we drove to Stagville Center and were told to follow a tour going to Horton Grove and then come back to the center. This we did, and toured the old slave quarters and barn, which are in the process of renovation and will be restored to their original integrity as funds are made available. The important thing, in my mind, is the fact that they are being preserved for generations to come.

The final leg of our tour was to Boo Hoo where the residential development, club, and golf course are located. I couldn't help remembering the many happy hours I spent riding Nancy in the Boo Hoo woods and hunting that area. Suddenly, it happened! I realized I was witnessing two pictures; one as a boy and the other as I saw it in 1990. The old Fairntosh and the new Treyburn. I could look at what has happened, what is happening, as an overview of yesterday and today, perhaps even tomorrow. I left with a sense of pride, knowing that that place, somehow, is a part of me and I of it.

Tobacco barn at Harris, 1938-39

During 1938-39, as a 14 or 15 year-old boy, I was charged with curing tobacco in this barn. My assignment was to keep the heat constant and elevate the heat in the barn as needed to dry the sap from the tobacco leaf stems. Once the leaf was devoid of sap, I was to "kill out" the barn full of tobacco.

It was a dreary job, but long hours of boredom caused me to begin reading dime novels about aviators in World War I which hooked me on flying. I usually had a pallet in the front yard of the barn and spent much time reading and dreaming about the romanticism of World War I combat flying.

Thus began my love for flying which transitioned into my role as a "fly boy" in World War II. I've always been glad I was able to live out my boyhood dreams.

SECTION II: WAR YEARS
MY GREAT WWII ADVENTURE BEGINS

My great World War II adventure began on March 20, 1943, at Camp Croft, South Carolina – near Spartanburg. I reported there to the induction center for processing and swearing in. Like so many others of my age, I was somewhat apprehensive and yet anxious to get into service. I was a student at North Carolina State, studying Aeronautical Engineering and struggling with my courses – not knowing when I might be called into the Army. Camp Croft settled all that. Mother and Dad had moved from Fairntosh in 1941 and were living in a small rural community north of Roxboro, North Carolina, where Dad taught vocational agriculture for a year (1942-1943). I was given a ten-day furlough to visit them and get my affairs in order before reporting to Ft. Bragg for assignment.

On the way from South Carolina to North Carolina, we stopped the bus at a "green front" store to buy some "red eye" to celebrate with those who had been inducted and drown our sorrows with those who hadn't. Above the town of Spencer, NC, we made a rest stop at a service station restaurant that boasted two bears – mama and papa. The devil made us do it, but we loaded papa with bourbon and 7-Up until he was bombed. Mama just stood there and looked at him in disgust as he wallowed on the ground – dead drunk!

In the Army Now

Upon reporting to Ft. Bragg after our furloughs expired, we were given the usual army physical, mental tests and shots called "processing." About the only thing that stands out in my mind about my week at Bragg is the food and a scare of spinal meningitis that quarantined our entire barracks for a couple of days. I also experienced the first of many G.I. parties. I realized that this wasn't going to be a picnic, but at the same time, it wasn't a bad life. Unlike most of my fellow recruits, I was not homesick. A year and a half at State College had pretty much gotten that out of my system. However, such was not the case with most of the fellows – especially those who were married.

Our barracks sergeant was an old, crusty individual who was colorful, loud and unforgettable. Being a farm boy accustomed to rising early, I was usually awake before reveille, just waiting for the scratched record to play the bugle call over the company PA system. "Sarge" would come in, switch on the lights, and, in his mildest, sweetest voice, would ask, "Did mommy's little boys get a good night's sleep? Are you guys comfy?" Then he would break into a line of profanity and yell like a bull, "Get your @$#%@&*sses out of that sack and be in formation in 15 minutes." Fifteen minutes, mind you, to dress, wash and make up the cot. We did it, too; have no doubt!

We learned quickly to do what we were told and to keep a low profile. That short week was in preparation for basic training at St. Petersburg, Florida. We were actually in a tent encampment – called Tent City – located on an abandoned resort site known as the Jungle, which had gone under during the great land speculation in Florida during the 1920s. It was a jungle, all right; no facilities and no comforts of home – nothing! Period!

We slept in six-man tents on army canvas cots and ate out of unclean mess kits since we had not yet learned to scour them with sand. The sand removed the grease left from washing the kits in greasy water from the GI cans. We also had a running bout with dysentery, flu, hunger, and all-round misery. And, we marched in close-order drill in ankle-deep sand all day. Boy was that fun! After a week of deprivation, the Surgeon General's office condemned the tent city, and we were moved into the old posh hotels of downtown St. Petersburg. Mine was the Princess Martha. We drilled on what was then the Yankee Baseball spring training field. I believe that field is now the site of the St. Pete convention center. I did not let it be known that I had prior military training at State because I didn't want to spend my war marching recruits at St. Petersburg, or anywhere else for that matter.

After about four weeks of this, we were ready for anything – so long as it was anything ELSE! I had been assigned at Bragg as a medical technician in the Army Air Corps. I supposed that would be as safe as you could get and was somewhat relieved. Perhaps this had something to do with having Aeronautical Engineering entered on my service record.

The army moves in mysterious ways. At any rate, that comfort did not last long. Near the end of basic training, several of us were called out of formation at breakfast mess and told to report to the dispensary in Class B uniform after chow for a 64 physical, better known as a flight physical.

Although I was long and skinny, I was, nevertheless, in good health and had no problem, except color blindness. The corporal who gave the eye exam asked me if I would like to fly and I said "Yes." I had always wanted to fly, even as a little boy, and the thought of becoming an aviator was too much to resist. The good corporal overlooked my color deficiency with the admonition that I would have to watch myself on future 64s. I did, and was caught only one other time – but that is getting ahead of my story.

Learning the Ropes

At any rate, those of us who passed the physical were shipped to Biloxi, Mississippi, for aircraft engineering training upon completing our basic training. Here I saw my first of thousands of B-24s, the plane that I would live with and fly and praise for the next two years. It sat in the hangar at Keesler Field, just outside Biloxi. Perhaps I should say that Biloxi is situated just outside Keesler. My classmates and I (class of 5-15-43) were to spend the long, hot months of the summer of 1943 in southern Mississippi studying every phase of the B-24 - structure, engines, frame, electrical system, hydraulics, test-stand, and on and on. The course was some 12 weeks long, and, when we finished, we were to be experts in aircraft mechanics.

Experts? By the time we reached our final weeks in the field, we considered ourselves "knowledgeable." As an exercise, we were challenged to disassemble a Brewster Buffalo or a P-36 and reconstruct it and pre-flight our subject. We used "A" frames to pull the engine and replace it prior to test flight. When my turn came, we wound up the engine with a starter dog (a crank to energize the starter), and I set the throttle to engage the engine. But, in my anxiety, I set the throttle too high. Upon engaging the engine, amid the noise and fumes, the engine coughed and roared into life. But for the fact that the tail assembly was tied to a pine tree by a three-fourths inch rope, I would have taken off in the densest Mississippi pine thicket imaginable. Needless to say, I quickly reduced power so the plane would release its strain on the rope, roll off the chocks and act like a good little bird. I flew later, but thank God that was the only time I almost flew a fighter solo through a pine forest.

We were constantly playing pranks on each other. One of our

"specialties" was to slip a hot washer from the negative pole in electricity into the baggy back pocket of a fellow student. He would be completely unaware of our mischief until someone slapped him on the rump. After much training in this exercise, we could slip a cold washer in his hip pocket and slap his derriere with great results. Later, we found that a simple pat on the behind would also have the same effect.

Another "goodie" was to catch two fellows sitting on an engine crate side-by-side and button the leg strap of one to the coverall leg of his partner (coveralls had straps and buttons on each leg above the cuff). Then on the pretense of finding something interesting, you would call one of them to investigate. When he jumped off the crate, he went headfirst and his companion went feet first. What a game! We short-sheeted cots in the barracks, tied shoestrings together, and did anything to break the monotony. Most of all, we had camaraderie. Our class began establishing friendships that have lasted through the years. We felt a closeness that many brothers do not feel. Some were lost in the war and we mourned them. Most of us returned and rejoiced.

Much More to Learn

Keesler was unique in that while we were learning the nomenclature of the B-24, we were also learning a great deal about ourselves. Most of us were in our late teens or very early twenties and had not been far from our home environments. We were a scatter of boys from all over the eastern part of the nation, thrown together in a common bondage – war. We learned that northerners and southerners basically enjoyed the same things and had the same aspirations. Our war brought us together as one people in one nation.

The summer of '43 was undoubtedly the hottest I have ever known – and the most humid. Drilling in temperatures of 100 degrees with humidity almost as high doesn't make for pleasant living. We attended school from 3 p.m. until 1 a.m., after which we were allowed to sleep until eight or so before call to reveille. Following breakfast, policing the barracks, washing clothes and drills, we had a few moments of free time before returning to class in the afternoon. Usually, we had Sunday passes to go into Biloxi, which was no big deal. A stroll on the beach, a game of bowling, girl watching (we could never get to know the local girls), or just goofing off at the barracks was our pastime on days off.

On several occasions, we were given 24-hour passes to New Orleans – a distance of two or three hours by bus. Since we attended classes until one in the morning, it was two or three o'clock before we could catch a bus that put us into New Orleans about daylight. There were all-night air-conditioned theaters that made excellent rest areas for a few hours to catch a little sleep before Bourbon Street called. Most of us were legally underage, but with the war on, no one seemed to pay attention to the drinking age. Besides, we often got free drinks from customers, bartenders or even sometimes from the girl performers. New Orleans was a good pass town, and most of us made the best of it.

General Ben Lear graced us with his presence on one occasion, and the base commander threw a parade for him. Of the 50 to 60 thousand troops stationed at Keesler – both basic trainees and aircraft mechanic students – perhaps ten thousand of us were mustered for parade on the concrete apron in front of the hangars, just after a July morning thunder shower. Naturally, General Lear's plane was late in arriving, and we stood in temperatures of near 100 degrees on that concrete apron in humidity beyond belief for nearly two hours. Over a third of the troops – and perhaps more – passed out from heat exhaustion and were carried off in droves in the "meat wagons." I stuck it out since I was appointed guidon that day and leaned on the squadron flagstaff for support. I was one of the lucky ones.

Our squadron adjutant was not pleased with the discipline during the parade, so the following day, we stood for another two hours at attention to learn how it was done. This adjutant was the same Lieutenant who told us frequently during squadron formation that a year from then, a third of us would be dead. The fact that our survival rate was much higher is perhaps due to our determination to prove him wrong. Likewise, I doubt that he would have lasted a year in combat. Someone would have shot him!

Then a famous "hot spot" on the beach known as "Gonorrhea Pier" was where the ladies of the evening reportedly plied their trade. It was a beer joint built in the Gulf several feet off the beach and was demolished by a hurricane that fall. A few of our guys went there on occasion. However, we were all warned to steer clear of the pier. As a matter of fact, Keesler is where we were introduced to the horrors of "social diseases." Sex education classes were held regularly, with lectures and films on what could happen as a result of a few moments of pleasure. Believe me, it was rough! My Dad had given me a lecture on cleanliness before I left home to go to college, but he didn't prepare me for anything like that. All barracks bulletin boards had a large poster showing an attractive young lady with the caption reading: "She May Look Clean BUT---."

During our last week at Keesler, perhaps September of '43, we were attending class in the field under primitive conditions. We slept in pup tents, ate out of the field mess and pulled mechanical repairs on our planes. This is the area I referred earlier in my experience when the fighter tried to take off through the pine trees. Everything went well for the first few days, until a hurricane hit the Gulf coast. It poured torrential rain all day in prelude to the actual hurricane hitting land. Our tents were flooded with four to six inches of water. The mess tents were held down by sheer manpower, and the food, when served, was floating in our mess kits. We stood all day among the towering pines, trying to present as small a target as possible. We did have raincoats and helmet liners, but these were little protection from the driving rains.

Then, about five o'clock in the afternoon, the full force of the hurricane hit with a fury. We got over five inches of rain in about an hour or less. Treetops were snapping off and flying with the hurricane winds. We didn't see them land; they just vanished. We were perhaps a mile from the beaches and caught virtually the full force of the storm, though I suppose the pine trees did lend some protection. Finally, after much deliberation and discussion among the officers, they decided that the risk was too great for the troops to remain in the field. We were gathered into formation and marched across the runways in shoe-top deep water back to our barracks area. Upon arriving in the squadron area, the eye of the hurricane passed over, and we stripped down to our shorts and went swimming in the streets. As you might surmise, I was elated when I shipped out of Keesler with a PFC stripe, headed for gunnery training at Harlingen, Texas.

Ah, Harlingen! After Biloxi, anything would look good. But, Harlingen, located in the lower Rio Grande valley, was a lovely community and an interesting base, as it still is. Today, the old air base is home of the Confederate Air Force. Our group arrived in Harlingen sometime late in September or early October of '43 to attend Aerial Gunnery School. We spent long hours learning the nomenclature and characteristics of the Browning Caliber 50 machine gun, the Martin upper turret, Sperry ball turret, Martin nose and Emerson tail turrets. We learned gun sights, deflection, mount movement, pursuit curve, lead, aircraft recognition and a host of other details germane to gunnery.

We ran several miles each morning in the pre-dawn darkness along the irrigation canals to keep in shape; stood retreat and studied each night for the next day's assignments. It was work and it was fun. I have always enjoyed handling guns, and this placed me in my element – shooting the caliber 30 and 50 machine guns, 03 Springfield rifle, Thompson machine gun and the Colt 45 pistol. We may have had other arms issued to shoot, but these are the ones I remember. My most enjoyable moments were on the skeet range. Having been an avid hunter, I grew up with a shotgun in my hand and had no problem. Others of my class were city boys who had never held a gun of any type. They were scared to death of the noise and recoil. I wound up as one of the top guns of gunnery school. All of this was referred to as "ground school" as we had not yet flown.

Becoming Airmen

Then came aerial training! My first flight was in an AT-6 "Texas," the backbone of gunnery training school, and the advanced trainer for pilots. It was a single-engine twin cockpit trainer and a great little ship. All my flight (eight of us, I believe) had taken their inaugural flight the afternoon before I did. When we arrived on the flight line the next morning, our pilot asked if anyone hadn't flown. I raised my hand, and he instructed my flight members to brief me on the procedure. Their main instruction was to fire from the left in short bursts. We were firing caliber 30s from a rear cockpit mount. The pilot took off with a very excited nineteen-year-old kid as his passenger.

My initial sensation was that of climbing a flight of stairs backward, since I was seated facing the tail of the airplane. As I was told to prepare to fire, I loaded the gun, charged it and swung left as I was instructed to do – MY LEFT – which was the right side of the aircraft. We suddenly went into a half roll leaving me suspended upside down, held in the plane only by a strap fastened to the floor, and my leg harness of the parachute. All this time the pilot was screaming, "Left, dammit, left!" I quickly got the idea and fired from the right side of the plane as soon as we were again in level flights. By this time, the sleeve target under tow from another AT-6 was so far away, I didn't have a prayer of hitting it. Upon landing, the pilot climbed out, stood on the wing and addressed the rest of the flight: "Are there any more of you damn idiots that can't tell the difference between a 'T-6 and a Jap Zero?"

Following the initial flight, I did reasonably well and qualified on all the arms with either sharpshooter or expert ratings. But, I carried the humiliation of not knowing left from right throughout gunnery training.

As we became more comfortable with our guns and planes, we were flown out over the Gulf for "splash target" practice. The base had poles driven into the shallow waters of the Gulf, and our training was to see where our bullets hit the water in relationship to the poles. This educated the gunner on mount movement – the velocity of the projectile in relation to the speed of the craft, which appeared to pull the projectiles ahead of you. Now, the fun came when flocks of ducks – by the thousands – flew underneath. With 200 rounds of caliber 30 ammunition in the magazines, we had a field day. We could see men in boats waiting to pick up the ducks on the water. I often wondered if any of those people were ever hit.

That fall, the entire duck population of the U.S. seemed to be gathered at our base. Occasionally, we would have to fly in circles over the base for 20-30 minutes waiting for a hole in the duck pattern before we could land. Our aerial training was all done at an auxiliary field; miles from the main Harlingen base. Having been cursed with asthma during my earlier years, that prairie with its pollen was no place for a person with any sort of a lung disorder. Several nights while on the range, I would suffer an asthma attack and have to get up and sit on the front stoop of the barracks, looking at the moonscape and listening to the coyotes howl. I knew that if I reported to sick bay, I would be grounded, so I suffered in silence while finishing my training. As soon as we left the auxiliary base, my asthma attacks stopped.

All of us enjoyed Harlingen. It was the first opportunity we had to really feel that we were becoming airmen. We not only flew the 'T-6, but the 'T-11 and Lockheed Hudson as well. We may have had one final flight in a B-24, but I'm not sure. At any rate, we graduated from Harlingen Gunnery School in mid November with gunner's wings and sergeant stripe, and a 30-day delay en route to our next assignment at Hammer Field in Fresno, California.

California Bound

I shipped my flight gear by rail to Fresno and caught a train from Harlingen to Winston-Salem to visit Mother and Dad. Fortunately, my brother, V.C. also had a furlough at that time, and we had a week or ten days together before he had to report to Camp Rucker, Alabama. Putting him on the train one night was the hardest part of the war. We each knew that we were going into the final phase of training before being shipped overseas and didn't know when or if we would ever see each other again. I spent the rest of my furlough with my parents and was fortunate enough to be there for Thanksgiving.

CREWING UP

In mid-December, I caught a train to Fresno. I attempted to hitch a ride on a bomber from both the Air Safety Command in Winston-Salem and the field at Greensboro but to no avail. The train ride took five days and nights to cross the country in old railroad equipment, some built long before the turn of the century. From Greensboro, I went to New Orleans on Southern Railway and then caught the Southern Pacific to Houston. In Houston, we transferred to a wooden coach of ancient vintage, taking us across Texas, New Mexico, Arizona and California. We changed in Los Angeles for Fresno.

As was the case with most other camp arrivals, we hit Fresno in the wee hours of the morning with no military transportation to take us from the train station to the base. Most of our movements were by troop train, but on this occasion, since we were reporting from furlough, we were traveling "TPA" (Travel Private Authority). Even though we only got a couple of hours sleep, we were awakened at normal reveille for processing.

One of my first priorities was to get my gear I had forwarded from Harlingen. When I arrived at the depot to claim my bag, the agent couldn't find it. We finally located my shipping tag on another contingent's shipment. My bag was lost with all my flight gear inside. The agent took me to a warehouse that must have embraced 20,000 square feet of space completely filled with barracks bags, flight bags, duffle bags and whatever the well-traveled soldier used for equipment. Literally hundreds of thousands of assorted bags were stacked from floor to ceiling. As we walked through the corridors of bags, near the rear of the warehouse, I paused and said, "Since the tag is gone, the only identification on my bag is a laundry mark T0263, just like that one right there." As I pointed, I realized that bag was marked T0263! My lucky day!

Temporary Assignment

The purpose of being temporarily assigned to Hammer Field was to "crew up" – that is, to assign the ten men to a crew who would go through phase training together and then into combat. My other crewmembers were: Wallace Clay, pilot, from New Orleans; Jim Bruner, co-pilot, from Wisconsin, I believe; George Burke, navigator, Dixon, Illinois; Paul Beam, bombardier, Evington, Virginia. The enlisted crew consisted of: Leonard Long, first engineer and top turret gunner from Lowell, Massachusetts; Willie Ray, radio operator from Riverview, Alabama; me as ball gunner-engineer; Bill Somerville, waist gunner-engineer from La Crescenta, California; Ed Burnett, armorer-waist gunner from Cincinnati, Ohio; and George Berens, tail gunner from Neola, Iowa. A more diverse group of all-Americans could not have been picked – ten men from ten different states: north, south, east and west. Our ages ranged from 20 to 35, with me being the youngest and Somerville being the oldest. Initially, we had a young Mexican as waist gunner, but he didn't work well with the other crewmembers and was mortally afraid of flying, so he was replaced by Somerville.

Phase Training

We then shipped to Muroc Air Base (now Edwards Air Force Base) and began our phase training. We flew countless hours over the California desert, Sierra Nevadas, Pacific Ocean, Nevada and Arizona. We flew bombing missions, gunnery missions and navigational missions. In the early months of 1944, we learned more about our B-24s and our on-board equipment – and we learned to endure the discomforts of cold high-altitude flying on oxygen. We learned bailout and ditching procedures and practiced over and over (on the ground) until we could evacuate the ship in something like 20 seconds – either up or down.

While none of us became life-long friends, we nevertheless functioned as a tight team and performed well, realizing that we were totally dependent on each other for survival in combat situations. Some of our more interesting flights took us over the Sierras at mountain-top heights, roaring along the peaks watching the animal life that few are fortunate enough to see – bear, moose, elk, deer, sheep and others. The scenery was spectacular. Then, too, we flew through Death Valley at below sea level altitude at perhaps 200 miles per hour. No doubt, the most lasting scene in my memory was sundown on the north rim of the Grand Canyon. Even with my color deficiency, I well remember the vivid reds and blues, yellows and browns on the rim of the canyon as it caught the rays of the setting sun.

Recreation

Our recreation was provided by passes into Los Angeles and San Diego. A friend by the name of Seay from Tennessee and I would catch a 6-by truck out of the Mojave Desert, through Lancaster, Mint Canyon, San Fernando Valley to Santa Monica Beach where the army had lodging. Mostly, we would go into Hollywood and Beverly Hills to look around and gawk like most soldiers do who have nothing in particular in mind. We did take advantage of the CBS Studio shows – Sammy Kay's "So You Want to Lead a Band," the Red Skelton Show, and the Lux Video program with Frank Sinatra. Ken Murray's "Blackout" of '43 and '44 was a great show, as was Earl Carroll's Theater-Restaurant. Although I didn't dance, we would go to the Palladium to listen to the music of Harry James. The nightlife of Hollywood was spectacular to a kid from North Carolina who had never seen such glamour. Oh yes, and we had to stand on the corner of Hollywood and Vine and watch the girls go by. But then, we did that in North Carolina too!

My two trips to San Diego were memorable because while there, I visited my cousins Ray and Virginia Calhoun. Ray was the skipper of a WWI destroyer, the Lambert, and was pulling patrol and minesweeping duty on the Pacific coast while a battle wound from a prior engagement in the Pacific was healing. He took me aboard the Lambert on one occasion, and I learned what it might be like to have one's own ship and to command a destroyer. Virginia and I sat up for long hours discussing family and bringing each other up to date on the news back home. While I was stationed in Fresno, Virginia and Ray drove up for a visit one evening. All the guys in my barracks were duly impressed that I was visited by a Navy Commander. Seeing kin from home in the distant lands of California was a real morale booster.

Another Physical

Just before completing our phase training, we took another physical that was almost my downfall. While attempting to "fill the bottle" in the latrine, two WACS/aids were standing outside the door hollering, "Make it quick, fill it up, you don't have time to play!" My bladder turned bashful, and I was a long time completing the assignment – so much so, in fact, that the other crewmembers were well ahead of me. By the time I arrived in the Flight Surgeon's office for the eye exam and color blindness test, I was alone. Since basic training and my initial warning, I had always memorized the yarn balls by watching those in front of me call off the colors. No problem, until now. I flunked! Not only could I not tell the good surgeon the yarn colors, he had me dead to rights when the dot charts were presented. Only by the persuasive efforts of Clay and Burke, as well as myself, were we able to convince the Flight Surgeon that my color blindness would not constitute a problem. After all, nine other pairs of eyes would be aboard the plane. At any rate, he let me go, and I was never caught again, otherwise I would have been grounded.

MY FIRST ASSIGNMENT

In early April, we were shipped to Hamilton Air Force Base in San Francisco to pick up our B-24 H that we were to fly in our combat assignment, wherever that might be. We spent a few days flying around the San Francisco Bay area getting accustomed to our new plane and gawking over the Golden Gate Bridge and Alcatraz. We named our new ship "Clay's Pigeon" in honor of our pilot; a play on the name and in defiance of the odds.

We were assigned side arms, parachute backpacks with jungle survival equipment inside, and we just knew we were going to the Pacific. It was logical - with the jungle equipment and the fact that we were on the west coast, nearer to the Pacific War than Europe - it all made sense. Whoever said anything in the army has to make sense?

Orders Received

We then flew cross-country from San Francisco to overnight stops in Sky Harbor, Phoenix; Midland, Texas; Warner Robbins, Macon, Georgia; and Morrison Field, West Palm Beach, Florida. We languished in West Palm Beach, in quarantine, for nearly a week while Burke was getting over some sort of virus.

Easter Monday morning found us winging south on a heading to an undisclosed destination. We could open our sealed orders an hour out of Morrison Field. Orders were opened, and we were on our way to the United Kingdom to fight with the Eighth Air Force. This was both exciting and scary. To be with the biggest air force and to go up against the Luftwaffe at the same time was certainly something to give us mixed emotions. Our first stopover was Waller Field, Trinidad and then to Belem, Brazil. Belem is astride the equator and at the southern mouth of the Amazon. The rain forest was impenetrable, as I found out when I went to the perimeter of the camp to see where the noise of drums was coming from. A blacker interior I have never seen!

The next morning we took off for Fortaleza, Brazil, and flew low over the rain forest. You certainly did not want to go down there. Our sister ship did the same thing and caught a big bird on its wingtip, forcing it to lay over at Fortaleza until a new tip could be flown in and attached.

A New World

We arrived in Fortaleza sometime after three o'clock in the afternoon and were met by a weapons carrier, and we were told to hurry and get our tent assignments before the rains started at 4:06. Although the weather was hot and extremely muggy, the clouds were billowy and in no way threatened rain. Reaching our tent in due time, we saw the field workers and Indians carrying their tools on their shoulders and heading for shelter. Then it came. Rain at 4:06, to the minute, in such a deluge that I have never seen before or since. Rain poured torrents for exactly two hours then stopped as suddenly as it started.

Every bird, every bug, and every animal native to that part of the world screamed, hollered, chattered, chirped, whistled, and croaked throughout the night. At sunup the next morning, everything was deathly quiet and remained so until after the rains the following afternoon. This went on for the next three days while we were billeted there pulling our 50-hour inspection on our plane. The rains were explained to me years later by a geography professor as a phenomenon peculiar to that area of the world where the sun rises and sets at the same time daily, and the trade winds of the Gulf Stream are constant. The clouds gather exactly the same amount of water each day, are carried at exactly the same speed, rise and convect at the same altitude, and dump their moisture at the same place. Weird but true.

Our 50-hour inspection completed, we took off early one April evening for the hop across the Atlantic to Dakar – then French West Africa. We had a twelve or fourteen hour flight, as I recall, and we played poker in the aft section of our bomber. We had bought a stalk of bananas and hung it from the overhead bulkhead so we could eat our fill of bananas. Those and K Rations were our snacks across the ocean. We also had extra bomb bay fuel tanks for the crossing and had racks in the bomb bay to carry our personal gear. Burke did a good job of navigating, for we arrived at Dakar with a small margin of fuel remaining.

Our night in Dakar was a memorable one for me. I saw a snake charmer, coaxing his cobra out of the basket with a small flute. He would slap the snake, and it would drop back into the basket only to be charmed back out time after time. Our canvas cots were filthy and the head roll and "biscuits" (the English version of a mattress divided into three sections) were even worse. I decided to sleep back in the plane and set out on a beautiful moonlit night to stand guard. I had my .45 and canteen, so I had no worries – until I reached the ship. Standing there was the tallest, skinniest, blackest Zulu warrior imaginable. He had on British shorts and carried a French rifle that looked to be about seven feet from butt to bayonet. Oh yes, and he had on a bush jacket and fez. He hollered "Halt!" which I promptly did. You don't argue with a fellow with a seven-foot rifle. I told him that I was standing guard at my plane, and he grinned, so I took a step. Again, he hollered, "Halt!" and again I explained my business. Well, after the third time around in this routine, I finally realized that the only word of English he understood was "Halt." I made a hasty and undignified retreat, back up the hill to the tent area and slept on those dirty bedrolls.

Dakar was the only occasion I ever landed on a mat runway made from corrugated steel. The landing made such a noise that I thought we had made a "wheels-up" landing. Also, wrecked planes were everywhere – British, American and German.

Our next stop was Marrakesh, Morocco. Our flight to Marrakesh took us over the Sahara Desert and through a beautiful pass in the Atlas Mountains. The pass was fourteen thousand feet and extremely narrow allowing no room for turnarounds. Once you committed to the pass, you were really committed. The floor of the pass was green and lush with a sparse dotting of small fir trees. Sheep grazed by a small shepherd's cabin nearby. It was a beautiful, pastoral scene, hardly recognizable as a path to war and destruction. I have thought of that beautiful pass many times in the ensuing years and of the incongruity of its location.

Marrakesh was a different world. As we arrived, we could see dozens of B-29s parked on the perimeter, heading for India, China and the Pacific Atolls. On that afternoon, we knew that Japan's days were numbered.

The base had an ancient temple of some sorts on it. It was made of stone and consisted of perhaps three or four rooms stacked one on top of the other with steps going up to each level from the outside. The whole thing overlooked a large, rectangular structure that probably had been a reflection pool or bath, but was overgrown and filled with shrubbery and brush. Here, prisoners (Italian) were encamped within barbed wire and stone fences with glass shards embedded in concrete around the top. They were prisoners of the French and were treated as sub-humans. They were fed scraps from our mess and wore only a few rags for clothing. Although forbidden, we threw food and articles of clothing over the fence to them.

Crews were quarantined upon arrival. Once our quarantine was lifted, we were issued passes for the afternoon of our second day. With a short walk to the heart of Marrakesh, we stepped into a different world. Filth and poverty were beyond description. We stopped in a French Café for a drink (we could only drink beverages with a high alcohol content due to the contamination) and while sipping our beverages, a leper crawled in who had no hands and almost no face. He begged for a drink, which the bartender provided without cost. The leper drank his beverage and returned the glass to the counter, without comment. We left the café!

A wall separated the new city from the old. We were forbidden from trespassing in the old city. However, we could stand on the elevated sidewalk and see beyond the wall. It contained mud huts, palm trees, dirt and people. The people were as deprived as you could imagine. Squalor and filth were everywhere.

Shoeshine boys, pimps and hawkers abounded. You couldn't turn the corner without being solicited – and that included the girls who plied their trade outside the city in the grass. After seeing the sights, we decided to go to the military center, which was some kind of an old chalet or castle. It was beautiful with its tile walls and floors, arched doors and windows and very spacious. We were served Coca-Colas and cookies and invited back.

Lesson Learned

Early the following morning, we took off for our rendezvous with our destiny in England. We landed at Valley, Wales, where we were promptly relieved of "Clay's Pigeon." We didn't know when we named her and had the names of each crewmember painted on her, that she would be required to go through armament for additional armor plating. We never saw her again. "Clay's Pigeon" blew up on her third mission from a direct hit by a German 88mm shell.

We were transported to Stone for "pool replacements" and then shipped to Clonto, Ireland for escape and evasion training. Besides hearing the stories of the escapees and evadees, we spent long hours on aircraft recognition. We had to identify all allied and German aircraft from front, side and oblique positions in 1/100 of a second. This wasn't easy, but that was about all the time you had in combat. We became experts in a hurry.

Clonto was just across the lake from Belfast and was cold and raw in the April weather of the British Isles. We were there about a month before we were shipped to our base at Attlebridge. Somewhere, we had a mix up in orders. We were supposed to report to the 466th Bomb Group, but instead, we arrived at the 446th. A day or so before our arrival, the 446th had been visited by "intruders." These were B-24s that had been captured by the Germans, refitted and flown back with our returning formations until they arrived at the base. Once in the landing pattern, with all crewmembers at landing stations, the intruders cut loose with 20mm cannons and tried to blow our planes out of the sky. It was dirty, but effective. The barracks where we were billeted for the one night we stayed at the 446th was punctured with 20mm holes. A lesson not to be forgotten.

THE 466TH BOMB GROUP

Then we arrived at Attlebridge, the 466th Bomb Group, which was to be our home and base of operations for the next four months. About the only memory I have of our arrival and initial training was a welcome from Lord Haw Haw and Dirty Girty from Luxembourg Kelly One - Kelly Two radio station: "Men of the 466th Bomb Group, we welcome Lt. Meyer and his crew to the European theater. We look forward to receiving them in our prison camps and await their arrival." Lt. Meyer and his crew were, along with us, the first crews of replacements to arrive at the 466th. We listened to Kelly One - Kelly Two constantly, for they played good American swing music. It surely beat the BBC.

"Rated" and Ready

Our crew was assigned Warchamp, son of Warchant, as our ship. Like Clay's Pigeon, Warchamp was a B-24 H. She had lost her left wing tip to a direct hit of an 88mm shell that hadn't exploded and had to await a replacement. That was our first mission and her second. We decided against renaming or painting our names on her station positions. After Clay's Pigeon, somehow, it didn't seem important.

The 466th at Attlebridge was located some eight or ten miles west of Norwich in the beautiful farming section of East Anglia. Interspersed among the three runways, perimeters and revetments were acres and acres of small grain (which the English called corn), farm houses, and barns. Life went on as usual despite the noises of war. Total mobilization was in effect, so women tended and harvested the fields alongside the few men who were left.

Our base spread over several hundred acres - perhaps thousands. It covered a wide farm area and was home to 3,000 plus men and 50 airplanes. We had three squadrons and were assigned to the Second Air Division - all B-24s. The First and Third were B-17s. Life was non-regimented at the 466th. If you had a flight jacket, you were "rated" (flying status). No saluting, no formations, no usual military discipline. We were a group of men sent to Europe to do a job and didn't need motivation. Teamwork got the job done. I have often thought in amazement how a bunch of civilians could be taken out of the cities and off the farms and become instruments of destruction. First of all, it took a certain amount of training – programming, if you please. And then, it took the pride of not letting your fellow airmen down. We were good, and we knew it. But we also knew that the Luftwaffe was good. Who was better? That's what had to be decided during those summer months of 1944.

I marveled at the "old hands" - those who came with the unit a couple of months before we got to England. They were combat veterans who had withstood the "test." Our crew was yet untried, but anxious to get started. Again, we spent long hours making test flights, getting acquainted with Warchamp, and the English countryside that surrounded the airbase. Both points were extremely important in cases of emergency.

Attlebridge had been constructed as an RAF (Royal Air Force) base and was only occupied by the 466th a short time before our two crews of replacements arrived. The group flew its first mission on March 22 to Berlin, and my crew and I flew our first mission on May 24th - the group's 36th. By the time we became operational, many of the "old timers" were about to complete their missions. We were billeted in Quonset huts made of metal, half-barrel shaped buildings that would accommodate two crews of enlisted men (twelve) in a fairly comfortable manner. Our barracks was in an area that formerly housed WAAF (Women's Army Air Force) personnel - female members of the RAF. It was located a mile or more from the mess hall, briefing room and another mile or so across the field from the revetment that housed our ship. Each ship had its own revetment, and each was dispersed widely around the base to reduce the risk of great loss in case of enemy bombing or strafing attacks.

To repeat, the base was widely spread over the English countryside, and we were only one of dozens of Eighth Air Force bases scattered all over East Anglia. To say that England was a veritable fortress at that time would be to understate the fact. This, in conjunction with the invasion forces gathering in the Portsmouth area, made the British Isles the most heavily concentrated area of men and equipment in the world during or since World War II. It was unbelievable how a beautiful countryside scene such as Attlebridge could come to life in the wee hours of pre-dawn and send up thirty or more instruments of destruction; but that was the game.

When not flying a mission, our time was virtually our own. The only formation requirement was mail call, but we would have met that without being ordered anyway. We were to read the alert list for the next day's mission, by the pilot's name, at mail call. Our daily routine was equally as lax. Besides mail call, we would read, wash clothes, "shoot the bull", play poker, go to the flight line and watch returning planes from that day's mission – or whatever else we wanted to do.

In the event we were not called for a mission on the following day, we could go into Norwich for fish 'n chips, see a movie or an excellent play, visit the museum in the Norwich Castle, pub hop, or just stroll the streets of the town. Our gathering place was the Norwich Cathedral where army trucks dropped us off and picked us up. The Cathedral was undoubtedly the most spectacular edifice in Norwich - built in the 10th century.

Although this was during wartime, Haymarket Square proved to be an interesting place. Originally designed as a market place for farmers and craftsmen to show and sell their goods, war had changed many of the customs and products. The square had become a swap-shop as well as a produce market and meeting place for the townspeople. Around the square were pubs, a theater, sundry shops, and fish 'n chips shops. As for fish 'n chips, my most vivid recollection of this shop was its location in a two-story building that had its roof blown off in an air raid. The upstairs floor served as the roof for the first floor shop, which contained a large iron pot - perhaps ten gallons - of boiling oil in which the fish and potatoes were cooked separately. A newspaper was on one corner of the counter that held a mound of salt and another stack of newspapers that were used to serve the fish 'n chips. It wasn't exactly the most sanitary way of serving, but the meal was delicious.

Norwich had received heavy damage during the early part of the war and was occasionally struck by a "buzz bomb," but by and large, it had been spared the devastation of some of the other cities that were larger and more convenient to Germany. We could occasionally see the searchlights from Norwich seek out an enemy plane and frame it as it flew over and dropped its bombs. This was more in the nature of nuisance raids than actual damage. Norwich is a lovely old town whose history goes back to the Norman Conquest. As a matter of fact, it has ruins of Norman walls on the outskirts of town and some that run through the rear of the Cathedral. A small river, quite picturesque, meanders through the town and back of the

Cathedral and is a navigational artery for barge traffic and pleasure craft. Any description of the area is inadequate. However, this account gives a general idea of Norwich's uniqueness and charm.

Life in London

London was another world, at that time, as it still is today. Buzz bombs had begun to fall on London shortly after D-Day and rained destruction on the city with a vengeance. The bombs seemed almost to fly in formation as they approached the central part of the city. Early on, the anti-aircraft guns tried to shoot them down, but it was determined that falling shrapnel caused more casualties than allowing the bombs to fall and explode. At night, the searchlights picked up as many buzz bombs as possible and followed them to their final target. The guttural roar of a buzz bomb is not likely to be forgotten. The tension during the last few moments of its flight is almost unbearable. When the fuel runs out, a "pop-pop" sound is heard followed by silence as the bomb glides to its destruction. The British are stoic people. By the time I arrived in London in late June of '44, they had undergone five years of hell. Yet, they were survivors and kept a resolve that we in the U.S. never knew. I had two or three passes to London and was captivated by its size, its "business-as-usual" attitude and the destruction that was evident everywhere.

The Night Life

The nightlife was THE thing that brought the G.I.s to London. "Piccadilly commandos" (prostitutes) were on every corner with twenty in between. London was in complete blackout, making it difficult to see who or what approached you in the dark. Likewise, the ladies of the evening wanted a look at you. The come on was "Got a light, Yank?" Then in the covered light of a flaming Ronson, you could look each other over. If she liked what she saw, it was "Four pounds for a quickie, mate?" And the bargain was struck - or not. Earlier, before we arrived in England, we were told by those who had finished their missions and returned to the states that a chocolate bar or a pair of nylons was all that it took to turn a fair damsel's head. Not so! The price went up with the escalation of the war. (I should point out at this juncture that these observations are those of a spectator and not a participant - remember the sex hygiene lectures?)

Scotty's on Piccadilly was a favorite hangout and gathering place for the airmen. It was an old pub that had endured the war and was frequented by natives and service men alike. Another famous, at least to us, landmark was the Wind Mill, a theater that boasted some of the most lovely nudes in all London - all properly decorous. They were all statue-still, and the shadowed lighting allowed them to show up just enough - but not too much. The Wind Mill was written up in a stateside magazine after the war and was billed as the "artsy" nude show of London. That it was!

Beyond the seamy side of this great city was an elegant and cultural side that overshadowed the bad. A walking tour from Piccadilly would lead through St. James Park to Whitehall, Number 10 Downing Street, Big Ben, House of Parliament, Westminster Abbey, Buckingham Palace, and Regency Park. If the tour was extended, it would cover Hyde Park, the Marble Arch, Trafalgar Square and numerous museums, theaters and parks. All this was of great interest to a boy from North Carolina; however, the buzz bombs drove me back to East Anglia and Attlebridge. I decided that combat was safer than London.

English Countryside

My crew and I had recently crossed the "Big Pond" (Atlantic Ocean), spent a month in Ireland studying escape and evasion tactics and aircraft recognition, and now we were ready for the BIG show. We were assigned to the 466th Bomb Group at Attlebridge, eleven miles or so west of Norwich, in early May 1944. We were to fly our combat missions in B-24 Liberators with the 96th Combat Wing of the Second Air Division of the "Mighty Eighth Air Force."

Our first couple of weeks were spent flying around the English countryside, becoming acquainted with the "lay of the land" and our new bomber – "Warchant." We had settled into our barracks, six enlisted crewmen who shared quarters with another combat crew – 12 men in all. We were crowded in a Nissan hut, which resembled and enlarged version of a 55 gallon barrel cut in half, lengthwise. Our quarters had a door on each end, as well as a companion window. Although it was spring and we didn't need it, we had a small stove in the exact center of the barracks, which was heated with charcoal, when available.

Quarters

Our Nissan hut was one of a dozen or so fronting the country road with a large sugar beet field in the back. Our latrine was several doors away and had much to be desired. Accommodations consisted of a row of washbasins, showers and chemical toilets that were supposed to be emptied daily. We mostly washed our own socks and shorts, but the heavy clothing would be farmed out to the local ladies, for a fee. Our pay was substantial, considering the comparable pay for the ground troops of equal rank; thereby giving us enough money to "exist" with a degree of comfort. We had cots with three "biscuits" (canvas cover stuffed with excelsior), which had a tendency to separate as we slept. But the good news was we did have a bed, such as it was, out of the elements, which was unknown to the ground troops. Although we had no sheets or mattress covers, we were issued two wool blankets. I still have mine.

Officer quarters were somewhat nicer, but still nothing elaborate. Their barracks were perhaps a quarter-mile closer to the center of the base than ours. The mess hall, headquarter building, control tower, drying room and briefing room were in the center of base activities with three runways forming a lop-sided "A." The main runway was 5,000 feet long. Ground personnel lived scattered throughout the base according to their job assignment.

The Way It Was

What made the base arrangement odd, in addition to the facilities already described, is the fact that farm families lived and crops were grown interspersed throughout the base itself. When the British Air Force built the base early in the war, they built the facility with the least disturbance to the population as possible. With German submarine blockades in effect, food was scarce and every attempt was made to provide rations locally. Many of our personnel became well acquainted with the British and were invited into their homes to share meals. I was never afforded that hospitality, mainly because I did not care to go to the local pub for a "mild and bitter" or "blood and mud." I hung out mostly in and around the base.

A non-flying day was boring! We had no formations to meet, except mail call, where we received our mission orders for the following day. That not only was a drawing card, but also SOP (Standard Operating Procedure). When not on duty, we slept late, ate, went to the revetment where our aircraft was receiving attention, if it wasn't schedule for a mission with another crew, talked with the ground crew, or just sat around and "chewed the fat." By mid-afternoon, many of us would collect near the end of the runway to watch the planes returning from their mission. That in itself was a humbling experience. Aircraft with damage or wounded aboard were given first clearance; the rest of the formation to follow. If thirty-two planes went on a mission, and only thirty returned, we knew that twenty men were either killed or captured. A three percent mission loss was "acceptable."

Getting Around

Our major method of conveyance around the base was either by "six-by" trucks, weapons carriers, which were slightly converted passengers, or bicycles. Literally hundreds of bikes were on our base; mine was no exception. I had always wanted to own a bicycle, but, due to asthma as a child, I was never allowed one.

Early in my assignment to the base, I was in Norwich one afternoon and while awaiting transportation back to the base from the Cathedral, I encountered a gentleman who wanted to sell his 'cycle for "nine pounds-ten" (about 40 dollars). He had a legitimate registration certificate, so I bought the bike and cycled the miles back to the base, registered it and was never happier! I now had my own transportation. She was a beauty, complete with a magneto headlight that generated light when the rider dropped a small generating wheel on the front tire, a tire pump, a repair kit and a luggage rack. It was a three-speed with caliper brakes and was light as a feather. When I finished my missions and was transferred to another base, I had no trouble selling it for my purchase price.

Mobilization

Great Britain was totally mobilized during World War II. Every eligible, able-bodied person, regardless of sex, was assigned to the war effort – except for the very young and very old. Most were in the military, but many were farm hands, while others were factory workers, truck drivers, air raid wardens, in the medical profession, firemen, and so on. Food was very poor, and rationing was the order of the day. Clothing was simple and in short supply, as was most everything else. Cars and buses ran on coke gas, operated from a small trailer pulled behind the vehicle, which burned coke and turned it into gas for the engine to burn for power. It was not a solution to the gas shortage, but it did work, after a fashion.

R&R

We had little time for "R & R" but did manage to get into Norwich frequently, to enjoy movie houses, legitimate theaters, pubs and girls! Norwich was a quaint old town that had sustained severe damage in earlier bombings by the Luftwaffe. However, while I was there, it was fairly calm.

It boasted a sizable shopping arcade, hotels, a fort/museum, the Cathedral, a massive railroad station, and the market square. That's where the action took place! Even in wartime – or because of the war – trading was fierce around the market. Everything from clothes, household goods, food and hardware was bartered – and maybe a lot more.

We had permanent passes and could go into Norwich on days we weren't flying or scheduled for the next day. Trucks would take us from the main gate into Norwich and deposit us at the Cathedral; reversing the process at night. Upon leaving the base, it was mandatory to take a handful of contraceptives – or no pass! No matter whether you used them or not, or how many were left over from the past issue.

An Unforgettable Trip

My second trip to London was very memorable. Upon arrival at the Victoria train station, I was greeted with a low guttural roar of a V-1 rocket that Hitler had unleashed on London following the invasion of France. They were devastating! Each carried a 1,000-pound warhead and was controlled by a gyroscope (no pilot) and powered by a ramjet engine, which sat on top of the bomb/fuselage. They were launched from the Pas de Calais area of France for the short hop over the channel to London. We had been bombing their launch sites, which were known as "Noball " targets. On my second night of the pass, one came over, cut off with a "pop-pop" and descended into the Piccadilly area, where I was. I hit the side ditch, and that's where I met my London guide for the next day… literally in the ditch! She invited me to her home the following day (it was Sunday) for lunch after which she would show me London from a native's viewpoint.

A Tour of London

The next day I caught the "tube" to East Stratford some distance from the heart of London. I arrived in the neighborhood and was told to look for the "lorry." No one told me it was a motor pool! At any rate, I found the family home – a row house common in England – and following a very meager meal, we headed for downtown London, again, on the "tube." We got off at Charing Cross Station and were immediately in front of Trafalgar Square. From there, we went to Whitehall and Number 10 Downing Street, to Big Ben, Parliament, Westminster Abby and Buckingham Palace. We toured Hyde Park, the Marble Arch, St. James Park and other points of interest. Boy was I tired! Following all that activity, we stopped in a pub for a beer and fish and chips before boarding the subway for East Stratford.

By the time I had walked her home from the sub-station, the subways had stopped running. Never mind…I took a bus that carried me back to Piccadilly. All went well until we got to a bridge that had been bombed, and the bus couldn't cross. The rest of the way was on foot. I didn't reach my room in the Hyde Park Hotel until early dawn, all the while flinching from the armada of buzz bombs coming over in droves. What a harrowing experience!

I arrived at my room a dusky dawn (it began to get light around 4:30 a.m. – double British summer time), and as I was undressing to get a short rest, a V-1 rocket exploded in Hyde Park just a short distance from the hotel. Then I realized why the hotel had ample room on the upper floors. I donned my jacket and ran across the park to where the bomb had exploded in a summerhouse. The concussion had sucked the wall off an apartment complex across the park, and I could see people running down stairs in panic. Beds, kitchen tables, and bathtubs were visible as if it were a giant dollhouse. That morning, I boarded the train and traveled back to my base where it would be safer flying missions!

So far as getting out in the countryside to see the beauty of rural England, we did not have much opportunity to see it. Mostly, we saw the British Isles from the back of a "six-by" truck, covered with a tarpaulin and viewed over the tailgate or from the dirty window of a train. I was able to see much more of the country on two return trips years later, but sadly, it had become more urbanized and industrialized, losing much of its "old world" charm. I'm glad I had the opportunity to see that part of the world, but most memorable was the stoicism of the English people, enduring extremely difficult times.

THIRTY-TWO MISSIONS

In all, my crew flew 32 missions over France, Germany, Luxembourg, Holland, Belgium and Denmark. We primarily bombed targets in occupied France and Germany. Many of our targets were in the Pas de Calais area and in Normandy. The former was in an effort to destroy the V-l (buzz bomb) launch sites, and the latter was in support of our invasion landing troops and follow-up tactical missions to support the advancing armies after they broke out of the beachhead areas. In Germany, we bombed such targets as Berlin, Hanover, Brunswick, Munich (twice), Saarbrucken, Karlsruhe, Eisenach, Bremen and a couple of others. In total, we bombed 12 targets in Germany.

In France, we concentrated on coastal targets in Pas de Calais and Omaha Beach in Normandy. We bombed "noball" targets (later identified as the buzz bomb launch ramps) Caen, St. Lo, Port En Bessen, Falaise, Melun, Mulhouse and many others I have long since forgotten. Some of these missions were spectacular, some routine. The endurance test lasted from four to nine hours in minus 30 degree temperature at twenty thousand feet, cooped up in a turret with no place to hide. That meant seven, eight or nine hours on oxygen, wearing a heated suit (we had no pressurization), a "Mae West" life preserver, parachute harness, flack suit, mask, helmet and goggles. To let you know the intensity of the cold, we gunners also had silk gloves we wore inside our heated gloves. If bare skin touched metal at those temperatures, the skin would instantly freeze and stick to the metal. In the event of a malfunction, we could take off our heated gloves and, with the protection of the inner silk gloves, could attend to the problem.

Our first mission was to Melun, France, due south of Paris, to bomb a Luftwaffe fighter base. This was not a particularly outstanding trip, and we thought that maybe it would all be that way. Not so! Our second trip was to

Mulhouse, France, to hit a railway marshalling yard. I watched our bombs make a direct hit on a train pulling out of the yard. The bomb exploded between the engine and tender and sent the engine flying end over end down the track. Next came "Big B", the fighter assembly plant at Tutow just north of Berlin. We hit the target with good results and started flying north to the Baltic Sea area when we encountered heavy flack that knocked us all over the sky. I was able to salvage a piece of flack that tore through the roof of our plane and landed behind me. I thought surely that I had been hit! As it turned out, we were flying over Peenemunde, Germany's rocket development center for both the V-l and V-2 bombs.

NORMANDY INVASION – JUNE 6, 1944

Just being afraid would hardly describe it. We were scared to death! This was the "BIG DAY," the day the High Command, General Eisenhower and fate had chosen to begin the long march on "Fortress Europe." D-Day – Operation Overlord – The Longest Day: whatever name has been tagged to the Normandy Invasion in ensuing years, this was June 6, 1944.

Following an early breakfast, briefing and takeoff, we of the 466th Bomb Group, Eighth Air Force, were winging our way to the French Coast to fly "tactical support" for the invasion forces hitting Omaha Beach at dawn. The briefing was anything but routine. Our briefing officer assured us that the Luftwaffe would be there in force to meet us. We were also told that in the event of a malfunction of our plane, we were to ditch and catch a surface craft headed for the beach area. With that may planes flying southeast, no one was allowed to turn back. Someone asked the briefing officer what we were to do when we hit the beach. His reply: "You become infantry and fight." The next question: "Where do we get a gun?" The reply: "That's the least of your worries; there will be plenty of them lying on the beach." Scared? You bet your life!

All the "big wigs" were there along with war correspondents and, of course, the briefing officer. The briefing officer, sensing his place in history, concluded with a little pep talk that went: "Men, today you will make history. I would give anything to be able to go with you, but I am to serve ground duty." A lanky Tech Sergeant stood up and replied, "Hell, Major, you can take my place if you want to!" That broke the tension!

Actually, we knew of the coming invasion the late afternoon of June 5th when all leaves had been cancelled, personnel were restricted to base, and we were ordered to draw our side arms and ammunition from armament. I picked up my .45 and at least fifty rounds of ammunition, just in case. Tension and excitement were high. We were called to mess around midnight, followed by briefing, then to the equipment room to dress in our flight gear. We were then transported to the planes, which had already been taxied onto the perimeter inline for takeoff. This procedure was most unusual as it hadn't been done before and wasn't done afterward. We were scheduled for a 2:30 a.m. takeoff in a drizzle and heavy overcast. We took off at thirty second intervals and circled left as we climbed, forming on our lead ship as he shot flares from a "very pistol" containing the group identification colors. Making formation in the dark sky among hundreds of other planes can be tricky, but fortunately the sky above the clouds was bathed in moonlight so forming up took about two hours and was performed without incident.

Upon forming, we set a heading for London by the time dawn was just beginning to break. The clouds began to scatter underneath, and we could see a solid string of lights from East Anglia, east of London to Selsey Bill on the southern coast of England. Barrage balloons were flying over London in great numbers to protect the city from low flying attacks. Shortly after leaving the coast, we began to sight the armada of ships and landing craft en route to Normandy. I had a commanding view as I was assigned to fly nose gunner-togglier for that mission – the first of many.

Thousands of landing craft cleaved the water 20,000 feet below us. All symmetrical, all in line, all craft were heading southeast to Omaha, Utah, Gold, Sword, or Juno beaches. The first two names were codes for the American Forces' landing; the second three were British, Canadian, and French. Moreover, as far as the eye could see: Airplanes! A thousand were in front of our group and as many behind.

I was struck by the sudden thought that I was witnessing the greatest spectacle yet devised by the human mind. The sheer logistics of putting that many boats loaded with troops, munitions, vehicles, and all the other gear that goes into an invasion of that magnitude was awesome. It was hard to believe from my bird's eye view that the Germans, and especially the Luftwaffe, did not yet know of our armada, or that this was D-Day. As we approached the French coast, we could detect our battleships, cruisers and other "Dreadnaughts" firing into enemy gun emplacements along the beachhead.

Our bombardier was ordered to operate the bombsight just in case of a mix up in the flight. Normally, the bombardier flew the nose turret and dropped his bombs on signal from the lead ship. The lead ship dropped a smoke bomb to indicate "bombs away," and the nose gunner had a toggle switch in the turret he activated at that time. We dropped on Port en Bessen just a few minutes and a few hundred yards ahead of our landing forces – to keep the enemy's head down for a few minutes. Hopefully!!

As we swung inland and westward, we began breathing in deep sighs. We had made it, and no German aircraft rose to challenge us. As it was later revealed, only two Messerschmitts were in the area that day to give air support to the enemy – against our 2,000-odd heavy bombers – plus fighters and medium bombers.

We swung northwest over the Jersey Islands and took a heading back to England. It is impossible to describe – even after 60 years – the relief, the exhilaration, we felt. I think we began to sense then that the German war machine was not the invincible unit we had dreaded all those months. We, perhaps, realized that the war's end was only a matter of time. If not at that stage, we certainly realized it in a few weeks when the beachhead had been established, and our troops had broken the Wehrmacht lines at St. Lo.

Later, I received a letter from my father asking if I had seen the "Big Show." My reply: "Yes I saw the 'Big Show." It's still the biggest one I have ever seen!

Memorable Missions

I will not try to recite each mission individually; to start with, I couldn't - not after all these years. I will cover some of the more spectacular ones, however. For instance, in Brunswick we approached on a north to south heading and began our bomb run. Flack was as heavy as I had ever seen - wall-to-wall black. As we approached the target, a B-24 broke in half, just behind the wings, in front of and below our formation. Five or six parachutes opened, but it probably did no good, for they landed in a vortex of firestorm caused by the exploding bombs. Even if they lived through that, I suspected the population of Brunswick would show them no welcome.

And then there was Saarbrucken, oh yes, Saarbrucken. We were ordered to hit the marshalling yard and approached from a west to east direction. Upon reaching the initial point (IP) flack became heavy and accurate. Flack exploded over and under our ship, and we could hear the shrapnel raining down on the wings and fuselage like rocks on a tin roof. Moreover, we could hear it pinging around in the bomb bay among the bombs. I flew nose on that mission and tried to hide behind the gun sight, but it was too small. My memory recalls losing two planes over Saarbrucken; however, the record of the 466th only lists one. I do recall an explosion in front of our ship and seeing parts of a plane fly by under the right wing - or were they bodies? I just don't know. I do know that we took hits in two engines and had to feather number one and reduce power on number three. This reduced speed to the point that we dropped behind the formation and called for "little friends" (fighters) for protection. Boy, were they beautiful as they escorted us home!

On our return, we heard a real drama unfold. Seems that one of the ships (perhaps not in our formation) had developed a fire on the flight deck and the "bailout" bell was sounded. All crewmembers bailed out except the radio operator who was under the flight deck changing radio frequencies. When he got back on the flight deck, he reached for an extinguisher and, in short order, put out the oxygen fire. He then called the group commander and asked for instructions. Like a father talking to his child, the group leader instructed the radio operator how to operate the automatic pilot that had been engaged by the pilot before he left the ship. He was told - in no uncertain terms - to fly way off the right wing of the formation and follow it back to the base.

Upon arriving at the base, the radio operator had gained considerable courage and decided to try to land his craft. "Not so," cried the base commander, who was in the control tower. The good radio operator insisted he could land it. The base commander insisted that he couldn't and would face a court martial if he tried it – even if he succeeded! The radio operator headed the plane out to the North Sea and jumped with two parachutes. Apparently, he discovered an extra aboard which he snapped to the seat ring of his parachute harness. He was the hero of the base, with or without his airplane. And, we imagined what the other members of the crew thought as they floated down in their chutes, watching their plane fly gracefully out of sight.

I witnessed several other oddities on different missions like a plane flying off our left wing taking a direct hit in the forward section. The plane did a complete slow roll, leveled off and exploded. While this was happening, a parachute was caught on the tail assembly with a body hanging on the end of it. The parachute slipped off and opened, but we surmised that the occupant had little chance of survival. On another occasion, en route to Munich I think, I saw a flight of ME 109s line up to attack a formation in front of us, but before they could press their attack, a flight of P-51s hit them and shot them all down. The last '109 was a first line fighter, and this incident attested to the lack of training the young Luftwaffe pilots had by comparison to our own. Then, there is the unexplained; a lone JU-88 attacked one of our sister groups on another mission, and made three passes from three o'clock high. He didn't shoot down a plane, nor did any of the B-24s get him. What a strange war!

From Ball Gunner to Nose Gunner-Togglier

My original assignment on the crew was ball gunner and back-up engineer (we had three on the ship), however, near the mid-point of our missions, our bombardier was removed from the crew so I became the nose gunner-togglier for the last sixteen missions. Those, plus the D-Day flight gave me a total of seventeen nose turret flights. It was an interesting place to be, but lonely. I could only exit the turret with the help of the navigator whose station was directly behind mine and beneath the pilot/copilot. Our escape, in the event we were downed, was through the nose wheel well.

No one on the craft had a more commanding view of what went on than did I. Too good in fact! I gave up hope for survival on at least two occasions – Brunswick and Saarbrucken. The flack was so intense that I didn't see how anyone could survive. Oddly enough, none of our crew sustained any injuries throughout the 32 missions. While we were issued flack vests, consisting of shingles of 1-1/2 inch zinc alloy plates covered with canvas, we opted to wrap our parachutes in them rather than wear such heavy equipment on long missions. As previously mentioned, we flew under extreme cold conditions, and the condensation from our oxygen masks would freeze and form great chunks of ice under our chins and down our flight suits. Occasionally, we would have to take the masks off and knock off the ice.

If I could get cigarettes to taste like those I smoked after each mission, once we lost altitude and took off our masks, I would be a chain smoker. I had the desire to inhale the whole cigarette in one deep breath. It felt good to relax a little and let our emotions drop. Likewise, after landing and parking our plane, we were herded to critique (de-briefing) for a session with Intelligence. Red Cross ladies would serve doughnuts and coffee, and the Flight Surgeon would dispense three fingers of bourbon. Now, that really did the trick! No food for half a day or longer, nerves a jangle and a shot of booze – wham! The purpose, of course, was to help us unwind, but, more importantly, it loosened our tongues for Intelligence.

The more a crew told, the more information was gleaned from the mission. Intelligence would then sort fact from fiction.

To say that we had nothing to eat on our long missions is not exactly accurate. We did have small lemon drops about the size of the last two knuckles of the little finger, which we could suck on by unsnapping the oxygen mask and slipping it in the mouth. Our lives completely depended upon those oxygen masks. One could only remain conscious a minute or so before passing out from the lack of oxygen at the rarified altitude of twenty thousand feet.

During our escape and evasion training, we were taught to make delayed jumps in the event our plane was shot down. First of all, if we opened our chutes as soon as we cleared the disabled plane, we might lose consciousness at a high altitude. Secondly, a falling body is less detectable than a parachute floating slowly to the ground. "How long should we free fall?" we would ask. "Until you can count chickens on the ground," was the reply. We were also required to wear our side arms on each mission; however, we were instructed to get rid of them upon reaching the ground. I suppose that made about as much sense as some of the other regulations we were called upon to honor.

We were relieved of all personal belongings before each mission, anything that would provide information as to the unit, hometown or individual's background. Our only identification carried in the plane was the dog tag. We were issued escape kits about the size of a paperback book that contained the necessary equipment to survive the first traumatic days of evasion. It contained a silk map of Europe, compass, money, chocolate concentrate, Benzedrine tablets (for a quick boost of energy), needle and thread and sundry lesser items. We laced our brogans (heavy work boots) to our parachute harness too – just in case it became necessary to walk. The well-dressed airman carried at least 50 pounds of extra equipment in flight, making movement extremely difficult.

To add to the burdens of the B-24, which was designed for a maximum takeoff load of 65,000 pounds, was all the extra gear, chaff (metallic foil strips that we threw out of our formation as we approached the target to jam the enemy radar), ammunition, fuel and bomb load. We often took off with loads in excess of 70,000 pounds. The '24 has a tricycle landing gear configuration with a tail skid added to protect the aft section of the plane. On several occasions, the tailskid would drag the ground as we taxied out for takeoff. Once on the end of the runway, under full takeoff power, the tail would lift into flight position. Then came the fun – getting all that equipment, fuel and explosives off the runway and remaining in one piece. Fortunately, we had one hell of a good pilot in Clay, and he never seemed excited – although we did clip an apple tree on a couple of occasions.

War is War

Since only nine of us completed our missions, we flew one man short; however, it was determined early in our tour that the ball turret created a 10-mph drag, and speed was more important than the extra firepower. Our formations were stair-stepped from low left to high right to give each other coverage both above and underneath the planes. Then too, we had excellent fighter coverage into, over and on the return from the target areas. P-47s would take us in, P-38s flew coverage over the target and P-51s would escort us back to the channel. In case of battle damage, a couple of them would fly protective coverage to prevent a Luftwaffe attack on a crippled ship. They would throttle back and lower flaps in order not to overshoot our slow, cumbersome bombers. Angels in the sky! When all went well on a mission, the fighter coverage would split and half of the element would "hit the deck" and strafe ground targets - munitions dumps, trains, convoys, flack towers, fuel deposits and any other likely target. Not even a bicycle was safe in Germany.

Keep in mind that most of us were kids in our early twenties who had done nothing of importance before the war. Few were professional soldiers and few remained in service after the war. This was a one-shot deal that brought us an experience that few had had before or since. The fact that we won the conflict in no way diminishes the dedication and superb training of the German soldier. He was good; make no mistake about it! But, for the idiocy of Hitler, our job would have been much harder. But then, if they hadn't placed the "little corporal" in power, there wouldn't have been a war in the first place. Had we not had the ability to destroy his factories and war-making facilities, his oil refineries and aircraft on the ground as well as in the air, I doubt that the tide would have turned in our favor as quickly as it did. Then too, we played Hitler's game. He bombed civilian targets in Great Britain, and we bombed the heart of Berlin, Cologne, Hamburg, Munich, and on and on and on.

We struck at Germany's will to make war as well as the factories and supply lines. While it was the Eighth Air Force's mission to fly daylight strategic missions, we flew many "tactical" missions in support of our ground troops when they were pinned down on the beaches of Normandy and after they successfully broke out at St. Lo. We dropped anti-personnel fragmentation bombs on the German troops defending St. Lo. We bombed them in the Faliase Gap where the American and British troops had them trapped. We bombed bridges, radar installations, and other targets to pin down the enemy and give our troops an edge.

Becoming the Lead Crew

As my crewmates and I neared the end of our tour of duty, our missions became few and far between. All the original crews who had initially flown with the 466th had long since completed their tours and gone home. Since we were the first replacement crews, we became the "old hands" and were subsequently assigned lead crew duty.

By late August of that year, the Allied troops had pushed the Germans back to the Rhine River and were collecting for the final assault on the Third Reich. Missions were longer as the lines pressed farther and farther eastward. Pathfinder crews were used more frequently as the late summer and fall weather began to deteriorate. Our crew was anxious to get the job done and get home, but the bomb group still needed seasoned lead crews. We fit that bill.

On our thirtieth mission, we were scheduled to bomb an aircraft factory at Oraineburg, Germany a few miles northwest of Berlin. We approached Germany from the North Sea with the route to take us in at the Denmark border, and from there, southeast to Oraineburg. We crossed the Zuider Zee and approached the Helogen Islands to find a wall of snow clouds reaching up to 35,000 feet - far too high for our bombers to reach. As we flew toward the mainland, a flight of B-17s was coming out over the cloud formation. Knowing we could not go over, under or through those clouds, a flight of three planes flew over the Helogens and drew flak, giving us credit for the mission.

Our last mission, September 5, 1944, was to Karlsruhe on the Rhine River. I flew engineer that day as we had a lead bombardier. My main job was to keep the fuel load balanced and everything tidy. We dropped bombs without incident, over heavy to medium flak, and turned for home for the last time. While we were not marginal in our fuel supply, we were running a bit low and the entire crew petitioned Clay to make a fuel stop in Paris, at Orley. What could be better than celebrating the completion of our missions than a night on the town in Paris? Clay would have none of it! His comment: "I've lifted this baby off the ground for the last time. When we land, we stay down." Nothing we could do or say would change his mind, and he had the controls.

I should mention that on our last seventeen missions, we were assigned a second B-24, "Slick Chick." Although she was battle worn, she was a great ship - and tough. This is the plane that we flew to Saarbrucken and lost engine power. We flew her back to the base with 76 holes in her, the hydraulic system shot out and trailing smoke from an oil fire on number three engine. We arrived over the base with less than a thousand feet of altitude left. Slick Chick flew 117 missions and was cited in the Liberty magazine as one of the more famous bombers of World War II.

Although I have only mentioned two aircraft, we flew many other B-24s. When Warchamp, or later, Slick Chick was under repair, we borrowed other craft. Spare Parts, Dumbo, Time's Awastin', the Falcon, Polaris and others whose names slip my memory. Warchamp later met her fate after I had finished my missions, but hadn't cleared the base. Her crew took her to North Germany and while flying at 20,000 feet, the pilot activated the automatic pilot which was set at 13,000 feet. She went into a dive and only by sheer strength and a whole lot of luck were the two pilots able to pull her out of it. When they did, the G-force pulled the bombs off the shackles through the unopened bomb bay doors and the landing gear sprung loose from their wheel wells. The skin along the sides of the fuselage was corrugated, and all the plexiglass and de-icer boots ripped out - but - she did make it back to the base! Warchamp never flew again.

Author's family in 1943: Front row: Father – Vestal Taylor; Mother – Lucy Taylor; Back row: Brother – Vestal Taylor, Jr.; Author – Glenn Taylor

Author – Staff Sergeant Glenn Taylor. Photo taken while enrolled in gunnery school in 1943. Author flew 32 combat missions with the 466th Bomb Group in the 8th Air Force out of Attlebridge, England.

Author and friend Martin enrolled in flight engineering school at Keesler Field, fall of 1943

Clay's B-24 crew - Back row L to R: Wallace Clay, pilot; Jim Bruner, co-pilot; George Burke, navigator; Paul Beam, bombardier; Leonard Long, engineer & top gunner Front row L to R: Edgar Burnett, right waist/armor gunner; Glenn Taylor, toggleer/nose gunner; George Berens, tail gunner; Bill Somerville, left waist gunner; Willie Ray, radio operator

Distinguished Flying Cross award ceremony officiated by Col. Fairchild. Receiving DFC L to R: Neish, Author, Somerville, Clary

Display of Author's service ribbons and medals, including dog tags and a piece of flack from a bombing run

RECOLLECTIONS

Before leaving the Attlebridge portion of this story, I would like to share a few random thoughts that have not already been covered. For instance, as already mentioned, we listened to Luxembourg Kelly One – Kelly Two radio for the best in popular swing music. The propaganda was relentless and was covered by "Lord Haw-Haw" and "Dirty Girty." He was English, and she was American. Both were working for the Nazi propaganda effort. Their "commercial" ran something like: "American airmen, do you realize that all your sacrifices are being made for Jewish political power directed from Moscow and Washington?" Both were arrested after the war and spent years in prison for lending aid to the enemy in time of war.

Sitting One Out

At five o'clock in the afternoon, we enter the mess hall for a twofold purpose: to get a fill of stew, and to check the alert chart for tomorrow's mission. As we pass through the door, we turn right and pause before the bulletin board. Here is the alert list: Myers, Everett, Morgan, Craft, and a few other names you are not familiar with. Each name is that of the First Pilot whose crew will be on tomorrow's mission. Clay's name (our pilot) is not listed, so we sigh with relief, knowing that tomorrow is a day of rest; we are not flying.

As the time passes and the sun recedes into a blanket of darkness, the base comes alive with activity. The ground crews are making last minute checks on their planes before the work begins tomorrow. Operations office is bustling with the arrival of the late orders from Division. The message center is busy giving and receiving coded messages. Armament and ordinance workers rush from ship to ship, loading the bombs, which will be dropped on "Festung Europe" tomorrow. Meanwhile, crews are sitting around smoking, playing poker, writing letters, reading, clowning, or "shooting the bull." No one seems particularly worried. Tomorrow is the time to worry. Tonight should be spent in complete relaxation. As we crawl between our blankets, we think how nice it is not to have to get up at two o'clock tomorrow morning.

Sometime during the dead of the night, we are suddenly awakened by the roar of trucks climbing the hill to pick up crewmembers alerted for today's mission. We hear shouts, some profanity, some jokes, and other shouts just to let us who are left behind know how lucky we are today. Another nap gives the crews time to "chow down," attend a briefing, go to the drying room to dress into their heated equipment, and catch trucks to their respective planes. As the first fingers of dawn push back the darkness covering the Anglican countryside, pilots begin pre-flighting their planes. Though we are a mile from the noise, we can detect every move in each cockpit by the sound of the engines. We hear a whirr as the starter catches and kicks each engine to life with a slight pause for engines to warm. We then hear the engine labor and decrease in sound as left and rights "mags" are checked respectively. Next both "mags" are off, and the engine sound decreases. Pilots then check the supercharger power "on" and "off" with the full power of the engine at 36 inches of mercury. Then the check is over for that engine. This is repeated three more times until all four engines are checked and ready to go. The planes taxi out of their revetments, down the perimeter to the runway, and take off one by one until all thirty or more are airborne. Takeoff is at thirty-second intervals and getting the group airborne takes fifteen minutes or more. We can then go back to sleep.

As the sun peeps through the crack between the window and the blackout curtains, shining directly into our faces, we awake. The first noise that strikes us is the sound above. The drone of thousands of bees circling overhead – only these are the sounds of a thousand "heavies." We get up, go wash, and cycle over to the mess hall for breakfast. Outside, the sky is filled with hundreds (maybe thousands) of tiny dots, followed by contrails as far as we can see. They have been forming for two hours now and are at an altitude of fourteen thousand feet, roughly. They fade from view as they approach Splasher 5 and head out over the North Sea on their way to Germany.

Back at the base, a peaceful day begins. This will be a seven or eight hour mission, with two hours already gone during forming. Now nine o'clock in the morning, the bombers should be returning around two o'clock this afternoon, at the earliest. Until about 1:30, activities at the base will proceed as normal. We kill time until noon chow, after which we cycle over to Operations for mail call and a one o'clock meeting of all crewmembers not flying. With the meeting over, we cycle out to our ship – providing it is not on a mission with another crew – and wait for the formation to return.

On schedule at two o'clock, we hear the familiar drone of approaching plane. Slowly the dots expand into a formation of airplanes. Thirty-six planes went out this morning; we begin to count them as they pass over the base in a triumphant "Buzz." Only thirty-five of them! Who got it today? We suddenly remember the two ships we saw go down yesterday: not a pretty sight.

One by one, 35 squeaks as the gear of each plane touches the runway. Each taxis into its revetment and "chocks" throttles, beginning a new deathly silence. We are all thinking, "Who was it that "got it," and how did it happen? Any escapees?

Suddenly, we hear a faint noise! Breathing stops to shut out any noise other than the engine sound. A distant dot develops into a majestic plane, badly beaten and limping in on three engines. What a thrill to see a plane and crew so determined in spirit! As it lands, unsteadily, it moves into its revetment to conclude the mission of the day. All planes are in and accounted for.

We walk away, mount our bikes and ride over to interrogation to talk with some of the returning crews. Everyone agrees that it was much easier than anticipated. We hear those comforting words, "You should have been there!"

While so near Operations, we decide to check the alert board for tomorrow's mission assignments. The list is already posted and, as we read it: Lawbrich, Cockey, Earnhart, five more names of no concern to us, and then CLAY. That's our crew. Tomorrow is not a day of leisure. Tomorrow we go back to work.

Celebrations

And then there was the 100th mission party held for the 466th in one of the large hangars on the base on August 18, 1944. Glenn Miller and his band played for those who wished to dance and beer flowed in endless quantity. Young English girls were invited to come to the party by the truckloads from every town and hamlet in the surrounding countryside. It was quite a party! Not being much of a dancer, I contented myself by just standing in front of the bandstand and listening to the beautiful music of the Miller band. His music had been my favorite since long before the war. I only had four more missions to go, at that time, and could imagine when I could return to the States and enjoy future band concerts in peacetime. Unfortunately, Miller was lost in a plane crash a few weeks later.

Another celebration took place on the evening of July Fourth. This one was spontaneous. Someone, somewhere on the base, fired a gun and was followed by another and another and so on. Rifles, pistols, machine guns, flares, sub-machine guns all joined in to celebrate the Fourth. Since the projectiles had to land somewhere, it was far from safe on the base that night. I took to the metal barracks, for whatever protection it might afford. There was an investigation; however, nothing came of it.

Losses

Our casualties were not all caused by enemy action. Planes sometimes crashed on takeoff with the loss of all lives aboard and sometimes collided in the air. The B-24 was not known for its maneuverability, which became more important in tight formations. Occasionally, a ship would get caught in the prop-wash of the formation in front and bounce over on top of another craft at the expense of two planes and twenty lives. I saw such an accident a few days after completing my missions. I didn't actually see the collision, for they were already at altitude, but while standing in chow line, I saw sheets of aluminum, tail assembly and other parts of the ship float to the ground. On first observance, I thought someone was burning cardboard boxes, and we were seeing ashes from the updraft.

Takeoff, under overloaded conditions, was undoubtedly the most critical, dangerous and frightening time of any flight – other than the actual bomb run. On one occasion, we had a ship that lost altitude and crashed, causing one of the 2000 pound GP bombs to explode, killing all aboard. The rest of us took off over the crash and proceeded on our mission. To make matters worse, upon returning from the mission and awaiting critique, the ground forces brought the pitiful remnants of a B-24 across the perimeter in front of us on a lowboy trailer. We protested the timing of such an act and were assured that it wouldn't happen again.

Although it may not seem important, statistically of the ten crewmembers who began our first mission together on May 24th, perhaps five or six of us completed every mission together. I recall that Clay, Bruner, Long, Somerville and I completed #32 together – and each one in between. I may have left out another – maybe Ray – but time does funny tricks with memory. The others either went on sick call, were on leave when a mission was called, or, as in the case of the bombardier, were taken off flight status. This leads me into another story – my alert for the 33rd mission!

Mishaps

Upon completing our 32nd, on September 5th, we went through the usual critique and afternoon mess before celebrating our completion. The flight Surgeon had been unusually generous in his bourbon allotment that afternoon before we caught the trucks to the mess hall. As we arrived at the mess, I jokingly said that I was going to check the alert list for the following day's mission (posted on the bulletin board at the entrance) to see if my name was listed. It was! I forgot about celebrating, food and everything else as I roared back over to the Squadron CO's office. I brushed by his aide and went storming into Major Elliott's office with a firm resolve to receive a court martial before I'd fly another mission. (I did remember to salute and address him as "Major, Sir.") His response was, "Let's check the record and see what has happened." As it developed, a spare gunner named Charles Taylor was supposed to be assigned to the mission and someone got my name instead. I thanked the major and all was forgiven – at least from my standpoint.

Some of the incidents were tragic and some were humorous. On one of our missions, "Slick Chick" was grounded for repair (it could have followed our Saarbrucken mission), and we were assigned another plane. As usual, the night was pitch-dark and the perimeter truck that transported crews to their ships dropped me off in front of a plane that was not scheduled to fly that day. I was challenged by "station defense" (a guard, one of dozens who protected the base). I heard him charge his rifle and yell "Halt! Mountain Home!" Mountain Home was the signal of the day, and I was supposed to counter with "Idaho." Do you think I could remember Idaho? No! In my most official voice, which I'm sure came out as a high pitched squeak, I said, "Just let me get back on the truck and go to the correct plane," which he did.

I believe that same evening one of our neighboring craft was being readied for its mission when the top gunner introduced his ammunition over the belt hold pawl and charged his gun twice, placing the round over the belt feed pawl and into chamber of the gun; whereupon the gun "ran away." It began firing on its own without the gunner depressing the trigger. In order to de-activate the firing mechanism, he had to grab the charge handle that kept the bolt out of the battery and stop the "run-away." This exercise takes a second or so for the thought process to tell the hand what to do. In the meantime, the gun had fired 20 or 30 rounds into the wing of the adjoining plane, penetrating the self-sealing wing tanks and setting the 100 octane fuel on fire (he probably had a tracer shell in there somewhere). At any rate, the ship was engulfed in flames, and to make matters worse, it was loaded with incendiaries that melted or exploded and added fuel to the fire. When we returned from our mission that afternoon and parked in the adjoining revetment, the only thing left of the airplane was the number three engine and the right landing strut. Everything else was just a pile of ashes in the outline of a B-24. A half million dollar ship gone! Punishment for the poor gunner was to read a short essay at the next day's briefing on "Why I should not charge the ammunition over the belt feed pawl on the ground." I suppose the "powers that be" decided that he wouldn't live long enough to pay for the craft out of a sergeant's pay.

Awards and Decorations

Awards and decorations came in profusion as we flew our missions. Air medals were issued after each five or six missions, and the Distinguished Flying Cross was awarded upon completing the tour. As a group, we were awarded the Presidential Unit Citation with an Oak Leaf Cluster for a couple of special missions that we flew. I don't even remember what those were for since it was a Group and not an individual citation. Likewise, we were authorized to wear the ETO (European Theater of Operations) ribbon and were awarded four bronze battle stars on it for major engagements flown in our tour as individuals. I do remember that the first star was for D-Day, one for Berlin and one for the Rhine River campaign. The fourth could have been for St. Lo or Falise Gap or something else long forgotten. I prize the four battle stars on the ETO ribbon above all the others. Not so incidentally, each award also carried five points toward discharge at the end of the war, and the DFC (Distinguished Flying Cross) added a few dollars a month to our paychecks.

Close Calls

Ted Clary was from Iowa and was the tail gunner of Lt. Meyer's crew with whom we shared the barracks. He was quite a character and a comic. On one of their missions Ted saw a "bogey" (enemy plane) approaching from the rear in the standard pursuit curve pattern. All crewmembers were instructed to report enemy aircraft by the clock position – 12 o'clock being the nose and six o'clock being the tail, and so on. Ted got excited and forgot all this "stuff." He called the co-pilot over the intercom with "enemy aircraft comin' in." To which the co-pilot questioned, "What time?" Ted's reply, "Right now, goddammit!!!"

In order to ensure safety, bombs were not armed with the detonator until after the bombs had been lifted into the bomb bay and hung on the bomb shackles. Armament assumed that these were safe until armed and would roll bombs off the back of a 6-by truck and let them drop perhaps four feet to the ground, then roll them under the fuselage and wench them into the bomb bay. On one such occasion, at the 466th, a bomb detonated on impact and wiped out plane, truck and ground crew in a flash. It was a 500 GP bomb, I believe. Why they were not using the bomb carts designed especially for this purpose was never determined, and no one was left alive to explain the inattention.

In the same light, once airborne, the armament gunner was charged with the responsibility of clipping the cotter pin that kept the armament vane (a small aluminum propeller that had to make several revolutions after the bombs were released) from spinning in the plane. Likewise, a wire fastened to the shackle was pulled out of the vane upon release, allowing a double backup safety. On one occasion, I was assisting the armament gunner in preparing the fuses, and as I followed him, I noticed, to my horror, that the vane on one of the bombs was slowly turning. We had clipped both wires by mistake. Having no idea how many revolutions the vane had turned or how active the bomb was, we carefully reinserted the pin and dropped it as a dud. I held the vane in my hand to keep it from revolving while the gunner re-pinned it.

Another close call came when one of the top bombs in the forward right bomb bay released before the bombs underneath. The ricochet effect bounced the bomb over the others and took the bomb bay door off its track and left it dangling below the ship. The rest of the bombs dropped without incident, but the door was hanging lower than the ground clearance of our ship. We had to climb out into the bomb bay and chop off the door with a fire ax, carried on all planes. This was done on our return over the channel without the use of a parachute. We couldn't use a 'chute due to space restrictions, and we could not cut the door until we let down low enough to not need oxygen. I have always had a fear of heights when it was oriented with the ground – a tall building or scaffold; but in the air, it never bothered me. Up there, I had no point of reference and was in a world all my own.

Tactics

The Germans were quite innovative in their attempts to camouflage their cities and factories. As a matter of fact, they would place netting over an entire city and camouflage adjoining countryside to look exactly like the real target, but miles away. The only thing wrong with this elaborate ruse was their inability to change the shadows as the day progressed. If the shadows were on the west side of the buildings in the afternoon, we knew that it was fake. They would even use smoke-pots to attempt to cover the fake target area. Our Pathfinders (planes using radar) would sort out the real target for our bomb run and MPI (mean point of impact). The Germans also used pink flack occasionally, whose purpose we guessed was to signal the next flack battery that our flight was getting beyond their range and the second battery should begin shooting at us.

I have already mentioned the "Intruder" incident at the 446th BG. We were constantly on the alert to a stray craft attempting to enter our formation and fly back to England with us for "protection." Once we heard but never saw a plane calling "W for Vilhelm, W for Vilhelm, come in please." This must have gone on for five minutes or more. Our radio operator opened his channel to "Command" so the entire crew could get a laugh. Finally, someone in our group in another ship replied, "Blow it out your barracks bag!" We didn't hear from that intruder again.

One of the more tense moments of the day at Attlebridge was in the afternoon when all personnel turned out to count the returning ships from the day's mission. First of all, we counted to see how many made it back; then we looked for straggler, and as they landed one by one, we witnessed the battle damage to the individual planes. How a craft could sustain such damage and still fly several hundred miles back to the base in England is a mystery. Moreover, how we could go out the next morning and do it all over again is even more mystifying. I suppose it has to do with training, discipline, courage, ignorance or just plain stupidity. Whatever it was, it kept the destruction of Hitler's "Fortress Europe" going until the surrender of Germany in 1945.

The Home Front in WWII

Realizing that the war in Europe would inevitably involve America, Congress passed the Selective Service Act of 1940. A song was immediately written: "Goodbye little dear, I'll be back in a year." But that proved to be a false prophesy when the Japanese bombed Pearl Harbor on December 7, 1941. The following day, Congress, at the President's request, declared war on Japan, a declaration that was immediately followed by Germany and Italy declaring war on the United States. We were at war!

The United States was just recovering from the worst depression ever recorded in our country's history. War meant instant jobs, but it also meant rationing, sacrifice, deprivation, and yes, casualties. America immediately tightened its collective belt and united in a massive war effort of proportions never before imagined.

Numerous agencies were formed to control war production, prices, transportation and a host of other "alphabet" regulations. Everything from soap to shoes; from gasoline to meat and butter to auto tires were placed under tight control. Gasoline, while still selling for twenty cents a gallon, was restricted to four gallons a week per customer. Automobiles were unavailable at any price. Ration cards were issued for assorted foods, clothing and other necessities.

Organizations were formed to collect vital materials to help the war effort. Civic clubs, scouts, college fraternities, church groups, school children, and others collected scrap metal, balls of tinfoil, (from cigarette wrappers and chewing gum), kitchen grease for cannon ammunition and a host of other materials. "Nothing was too good for our boys in the service."

As the volunteers and draftees left home to report to the hundreds of camps throughout America, wives and girlfriends banded together to wrap bandage and write their loved one letters of encouragement. These same ladies stepped forward to fill the production gap in industry and became known generically as "Rosie the Riveter." This would continue until the enemy surrendered. Able-bodied men, too old or not draftable, were also joining the aircraft industries and shipyards. No one was left out of the industrial circle.

Windows displayed blue stars for men in service and gold stars for those who lost their lives in combat – tragically over 300,000 by the end of the war. But the home front was resilient; with all the sacrifices and hardships, the population remained dedicated to final victory. Service men were invited into homes for Sunday dinner to add a "touch of home" while in the army.

Radios and newspapers were filled with news from the war front and the evening news was the most popular program on the radio. We didn't have television in those days, so much of the news information had to be described in detail or, in the case of the newspapers, maps of battle lines and engagements were drawn in detail.

Entertainment was limited to movies (usually with a war theme) or dancing at the local "nightspot" where girls jitterbugged to Glenn Miller's band – since many men weren't available. Concerts in the park and special celebrations, such as the Fourth of July and Armistice Day (later reclassified as Veterans' Day), were held.

The "folks back home" certainly had no walk in the park, but they were there to undergird those millions of troops that served and helped win World War II. These were certainly the "silent heroes," and the victory could not have been attained without them and their resolve. A great big salute to "the folks back home!"

MY TOUR IS COMPLETE

I had done it! I completed my tour and came out alive with a reasonable degree of sanity. The relief was indescribable. After a few days of processing, we broke up the crew and were shipped in all directions. I never saw any of them again, except Somerville who was shipped to the same casual pool with me at Attleboro, a small base just south of Norwich. Several dozen of us ex-combat airmen marked time at Attleboro, awaiting shipment back to the states.

By this time, it was October and the period of Montgomery's folly of "Operation Market Garden" – the attempt to land allied troops behind the enemy at Arnhem. That morning, we were awakened to the drone of thousands of planes overhead, C-47s, each towing two gliders loaded with troops. A few days later, we were assembled in the post theater and addressed by a Lieutenant who asked for volunteers to fly one more mission as flight engineers to drop supplies to the beleaguered troops at Arnhem. The supply ships, all B-24s, would fly in at low altitude and make their drop. Each of the participants was promised an extra stripe for the mission. In all, I recall eight volunteers. Only five returned. Personally, I was happy with my Staff Sergeant's stripes and had no inclination to run any undue risks, especially after having survived the regular tour of duty. At any rate, time moved slowly while we awaited transportation. Initially, we were to fly home in C-54s, but were later told that we would sail on the first available ship – which turned out to be the Queen Mary.

Steaming Home

We were transported by rail across England and up the west coast to Glasgow, Scotland, where the Mary awaited. Approximately 600 airmen, 600 Normandy casualties and a like number of army troops who were being rotated from Greenland gathered. We departed Glasgow Harbor in late October and crossed the Atlantic in five days, using a zigzag course through the south Atlantic and up through the Gulf Stream to New York. The purpose of the circuitous route was to sail in as calm water as possible for the comfort of the casualties aboard. These men were heartbreaking – some without arms or legs, others without faces or sight. They had a buddy system, though; a wheel chair case with sight was propelled by a blind soldier with arms and legs. The armless were fed by those with arms and so on. This exhibition of courage made the rest of us even more aware of our good fortune in surviving whole.

Those of us in the Air Corps pulled M.P. (Military Police) duty aboard ship while the Greenland troops pulled K.P. My assigned guard post was between number three and number five staircases on "C" Deck – through the enlisted men's mess, the kitchen and officers' mess. Being a British ship, we were not served the tastiest foods anyway, and we were hungry much of the time. The mess kitchen did bake delicious hard rolls for the officers' mess that were on display in the kitchen. I did what any red-blooded hungry kid would do and stuffed them inside my flight jacket for a late snack! The mess Petty Officer ordered me to "double the guard." I did and enjoyed the opportunity to take twice as many rolls!

The Queen Mary sailed without escort, as it was so fast that it was safer alone than sailing with a much slower escort or in convoy. It was lightly armed with nothing more than a few caliber 50 machine guns and 37mm cannons. It may have had a few heavier guns, but I didn't see them. As M.P.s with armbands, we had free run of the upper portion of the ship and explored virtually the entire area, except the engine room and the pilothouse. The main staircase was elegant with its sweeping rail, marble steps and mirrored walls – Art Deco at its finest. The odd thing was to stand in the midst of this splendor and watch it sway back and forth.

On our last night out of New York, we hit a hurricane that turned the Queen Mary every way but over, and sometimes I thought even that would happen. Waves of such size as to break over the entire ship were common. Of course, the ship had to be listing away from the wave to allow it to get a "running start." Keep in mind, however, that the Queen Mary stood some thirteen stories above the waterline and was in excess of 1,000 feet long. Occasionally, she would breach two waves, allowing the screws to come out of the water. When this happened, she would vibrate violently. Although the strength of the hurricane abated, the ocean waves remained awesome. We stood off New York's harbor all day waiting for the water to calm enough to bring the tender that held the harbor pilot alongside. Finally, in the late afternoon, he came aboard and brought us into the harbor, up the Jersey River to berth. As we passed the Statue of Liberty, all of us craned and stared at the symbol of that for which we had been fighting to protect. My helmet and gas mask are resting on the bed of that river somewhere off a point from the statue.

A Hero's Welcome

We were welcomed to Camp Shank, New Jersey, on October 28, 1944, with a short speech by the post commander, a steak dinner, apple pie and ice cream and all the milk we could drink. The next day, Sunday, we were given a pass into New York City to see the bright lights of "The Big Apple."

The following day, we were put aboard an express troop train (there were not many of those that carried priority) and sped overnight to Fort Bragg for processing, orientation and the issuance of a thirty-day pass. I caught a bus from Fayetteville to Greensboro and transferred to a Winston-Salem bus. For some strange reason, I had an almost uncontrollable desire to stay in Greensboro overnight. I don't think I was ready yet for the outside world as much as I wanted to see my family. I overcame that emotion, though, and arrived at the doorstep of 2050 Elizabeth Avenue, Winston-Salem, ready for a hero's welcome. I rang the doorbell, and Mother wouldn't come to the door. It was Halloween, and she thought I was some mischievous kid from the neighborhood. I finally got a response after several persistent rings and was welcomed with open arms. In our correspondence before leaving England, Mother had written that she had purchased an antique preserve bowl like Grandmother Chilton had. I wrote her that I wanted peach preserves in it when I got home. Sure enough, there sat the bowl full of peach preserves!

Dad was an inspector for a defense plant that made batteries for the Navy proximity fuse and was on an evening shift. After Mother called him and the aunts, uncles and cousins, we had a family reunion that night. Then, I felt that I had the welcome I was looking for. I did realize, however, that I had to rehearse my conversation before I spoke, else I would break forth in "barracks language" – which I almost did from time to time. I was glad to be back home, but it was also nerve wracking. Being among the first returnees from "the wars," people had thousands of questions. "How was this; how was that; what did you do when, and on, and on." I was out of my element and wasn't comfortable among civilians – even my own family. Once or twice, Mother and I went downtown with me wearing my uniform, complete with wings and ribbons. She would introduce me to an acquaintance and the response would be “And he has done so much and is so young.” This would invariably tear me up; after all I was 20! To avoid these scenes, I would put on the leather flight jacket – with bombs (32 of them) and a B-24 painted across the back. This created even more of a furor. As the saying goes: “Everyone likes attention,” but this much---?

My furlough days flew by, and shortly I was on the train for Miami Beach for rest and reassignment. I spent two glorious weeks there, living in the old posh hotels and eating food that would make a king blush. More physicals, evaluations, orientations and conferences took much of the days. Evenings were our own. My classification officer suggested cadet training. I had some college and nearly two years of air force training behind me, and the army still needed pilots. The only thing I would like to have been that I wasn’t already was a P-51 pilot. I knew my chances of making it through pilot training were slim since I had a color deficiency, so I opted for gunnery instructor training.

MY ADVENTURE CONTINUES

Arrival at Laredo, Texas, was not the highlight of my military career. My service record was stolen in the train station in Miami before departure. It isn't easy to explain such a careless act, and when I tried to, the captain replied, "You what?!" I had to languish at Laredo for several weeks to get my service record redrawn. About the only thing we had to go on in rebuilding my two years of service was my shot record and awards and decorations. But, the Adjutant-General's office came through and made me legal again. During that period, I only had a 3 x 5 card in the squadron index file to attest to the fact that I was in service.

Gunnery Instructor Training

Once the record was reconstructed, I started classes to become a gunnery instructor. They taught both ground and aerial class procedures, and much time was spent catching up on the new approaches to gunnery since I had gone through Harlingen over a year earlier. For instance, the army had developed the "jam-handy" and Waller trainers to hone the proficiency of trainees. The "jam-handy" was much like the shooting gallery at a train station – except much more sophisticated. The Waller trainer was a turret located in a quarter sphere with five projectors projecting a fighter slashing across the screen. The turret had an electric eye, and when the gunner-operator scored a hit on the projected screen, a bell would ring in the turret, and a meter would record the number of hits. This idea was later developed commercially and called Cinerama. We had tests in night blindness and aircraft recognition. The latter was more or less just an exercise, since our code overseas was "If the bastard points his nose at you, shoot him." There were too many "intruders" in combat to take a chance.

Then began ground gunnery. We shot from flexible hand-held gun mounts, turrets and trucks. The truck exercise was the most fascinating of the ranges. Pickups with high sides were used, traveling around a mile or so elliptical track with skeet houses along the way. There were a dozen or fifteen houses from low to high on a ten or twelve foot scaffold. The low houses were dugouts with a three or four foot construction that contained the release mechanism, the operator and the bird supply – and occasionally, a rattlesnake or two.

Everywhere were signs "BEWARE OF RATTLESNAKE." This is as frightening as the real thing since I am mortally afraid of snakes of any variety. One of our boys did encounter one in a high house under a box of skeet birds. Fortunately, it was a cold February morning and the snake was listless, but the operator wasn't. He ran about ten feet straight out of the high house door with nothing under him but air – then he began to fall. After first trying to use a shotgun, the range sergeant killed the snake with his .45, but only after the snake had been driven out of the close quarters. We were all extremely cautious after that episode.

Flight training was worse than combat. Our pilots were unseasoned, and takeoffs and landings were "white knuckle" events. One of our pilots stalled out on landing, but fortunately, he caught it just before the B-24 hit the ground and brought the plane over the runway for a safe but hairy landing. This condition was to prevail for the rest of my time in service until I took myself off flight status. Perhaps as many crews were lost in training as combat – maybe more. We shot sleeve targets (towed by AT-6s) and "Pinballs," which were P-63s that had a thin coating or armor plating for protection because the gunner shot directly at the plane, using plastic bullets. The plane was wired to record hits much as was done in the Waller trainer. Some twenty-five of us finished instructor training in March, 1945, and received the highest MOS (Military Occupation Service Rating) an enlisted man could rate in the air corps – a 938. This denoted that we were proficient in engineering, armament and gunnery and were qualified to instruct in aerial gunnery.

Border Runs

Laredo is an old Texas town that rests on the bank of the Rio Grande River and across stream from Nuevo Laredo, Mexico. It isn't a pretty town, but is somewhat quaint. During my stay there, I crossed into Mexico many times on shopping excursions. Oddly enough, goods were available in Nuevo Laredo that had long since either been rationed in the States or weren't available at any price. Things like appliances (name brands), nylons, good American whiskey, watches, clocks, etc., were available and reasonable. I sent a lot of this stuff back to the folks in Winston-Salem.

In Laredo was a unique bar called the Silver Dollar. It was divided by class distinction and included several rooms serving peons to the gentry. Being in uniform, we could go from room to room and visit with the natives at any of the several "watering holes." Some of my friends crossed the border one evening to buy some black market gasoline and stock up on whiskey. They were allowed a quart per person through customs; however, they had purchased much more than allowed, so they tried to hide it all. The border guards caught them and confiscated the entire lot. When they got back to the barracks in the wee hours of the morning, they woke the rest of us to force us to have a drink of their $100 Scotch. Seems that they had talked one of the guards out of one bottle.

The food at the mess hall was probably as bad as any I had in the army. Canned or powdered milk with cereal for breakfast, toast and powdered eggs were common. Lamb or mutton for lunch and dinner was a regularity. We got to the point that we would start bleating like sheep as we came over the hill approaching the mess hall – much to the anger of the mess officer. He must have gotten a heck of a deal on lamb – or a kickback?

In view of the food quality, or lack thereof, some of us would eat toast, butter and jam for breakfast, go to the PX for a light lunch and into town for dinner. We discovered a restaurant that served a pound and a half steak with all the trimmings for about a dollar sixty. Two of us would order and when that order came, we would order another steak to split between us. With that kind of fare, we didn't go very hungry.

We were warned to cross the border in pairs for protection. It was common to find a body floating in the Rio Grande with a stab wound in it – sometimes a G.I. who thought he was tough enough to look after himself and couldn't. I was glad when our training was over, and we met for reassignment.

Reassignment

Three stations were open for assignment: Charleston, South Carolina; Mountain Home, Idaho; and one other whose location I have forgotten. I do remember that it, too, was in the boondocks like Mountain Home. We drew numbers: one for Charleston, two for Mountain Home, and three for the other location. I drew Charleston, two hundred odd miles from home. I was offered a hundred dollars for my number.

I was a gunnery instructor at Charleston Air Base in B-24s until V.E. Day. Again, we were flying with green crews, and flying could get hairy. The six gunners were my responsibility while the plane was airborne. Ground instructors provided classroom instructions, which relieved the aerial instructors of a great deal of monotony. The crews were in phase training, similar to our training at Muroc, which consisted of long hours of flight, practice bombing and aerial gunnery. We often flew practice missions up the coast to Northern Virginia and dropped the bombs on the way back with a session of target practice on the way. It wasn't all that exciting, but it was a job.

Three of us (a pilot, a bombardier/navigator, and I) went into town one evening and took on a lot of joy juice. In our mellow mood, we pledged as one to volunteer for combat duty in the Pacific war on an A-26. We were tired of the monotony of the daily life and routing flying of Charleston and wanted to do something exciting. The next morning, although sober and contrite, none of us had enough sense to admit that the previous evening's pledge was a tragic mistake, so the three of us marched into the squadron commander's office and made our noble request. He looked at us like we had lost our minds (which I guess we had) and, in his most stern voice said, "Get the hell out of my office; I don't want to see any of you again." That was my last noble effort, and I felt a relief that is difficult to describe.

Charleston was a lovely city, and I enjoyed my three months there. My cousin Bob Taylor and his bride, Margaret, were living there. They were stationed at the base, and made life in Charleston more bearable. I have been back many times and always approach Charleston with nostalgia. On Easter Sunday of 1945, my parents traveled to Charleston by bus (there was still gasoline rationing) to spend the holiday with me. We attended Sunday service at the Citadel Chapel. Following the service as we were leaving the church, I heard a booming voice shout "Sergeant!" I turned around and found myself staring at four stars on each shoulder – General Summerall – Commandant of the Citadel. He proffered his hand and said, "Son, I can see by your ribbons that you have been in combat and returned safely. I know your parents are proud of you. If I can assist you in any way here at the Citadel, please let me know." With that, he spoke to Mother and Dad, turned and left. He was a GREAT man! I have never forgotten the kindness the General showed nor has my appreciation diminished.

Back on the flight line, we were getting crew assignments one morning when a fellow instructor asked to swap crews with me. Apparently, the copilot of the crew was a hometown friend of his, and they wanted to visit. That was fine with me. On returning from the day's practice mission, we were ordered to fly "spotter coverage" for a ship in distress. That ship was the one I had originally been assigned. The controls had frozen, and it was impossible to land it. We were flying at 10,000 feet over Summerville when the crew began bailing out. All eleven men cleared the ship in good order. However, Summerville is in the midst of a swampy area that is almost impenetrable. I know, because I hunted in that area in later years while working for Southern Railway. The crew got back to the base in ones and twos, battered, torn and bruised...but alive. The ship didn't fare as well. When the pilot jumped, the plane rolled up on one wing and plunged straight down to its destruction. We heard the explosion at our altitude over the roar of our plane's engines.

V.E. Day

When V.E. Day finally came, all personnel of the air base were restricted to base. Civilian and military authorities did not want to turn the Air Corps and the Navy loose on Charleston at the same time to celebrate. Our turn came the following night when the Navy was restricted to their base. Three or four of us were celebrating together, and we were adopted by a cab driver who was apparently still celebrating from the Navy's liberty the night before. At any rate, he provided transportation we needed, and refused to take a cent for fare. "Nothing is too good for the men in uniform." He even took us to his home, where a cold reception awaited from his wife and children. He claimed complete sobriety and his high station with the Charleston police. To prove his point, we drove up to the police station, and he rolled out on the sidewalk in front of the night captain. The officer simply said, "Tommy, Tommy, you again?"

V.E. Day virtually made our mission of training new crews obsolete. With the cessation of hostilities in Europe, the air power of the Eighth, Ninth, and Fifteenth Air Forces could now be directed toward Japan, making fresh B-24 crew trainees unnecessary. They did, however, still need medium bomber crews in the Pacific, so we were reassigned as gunnery instructors in B-25s. I was finally in my element.

In My Element

We were transferred to Greenville Army Air Base at Greenville, South Carolina, in late May. The B-25 was much smaller than the '24 and was much more maneuverable and faster. The '25 carried a crew of six or seven, depending upon its mission, and was heavily armed with caliber 50 machine guns – 13 in fact, nine of which were forward firing. Six of the nine were nose or package guns that were fired by the pilot in strafing or ground support missions. It was a low-level attack bomber that carried out a tremendous mission throughout the world and was in virtually every theater of operation. Again, as in the '24, my assignment was to teach aerial gunnery to the crewmembers. Although our base was at Greenville, our bomb and gunnery ranges were at Myrtle Beach.

Flight time from Greenville to Myrtle Beach was approximately one hour, or at most, an hour and a half. We would practice bombing secluded targets in the flat country of South Carolina and then shoot at sleeve or splash targets to acquaint our gunners with the sound and action of the plane, turret and ammunition. The splash targets were, as in Harlingen, Texas, poles driven into the sands of the Atlantic Ocean a few hundred feet off shore. We flew an elliptical pattern from Myrtle Beach over the Atlantic, south to Brook Green Gardens, then north in a wide right swing (east) across and over U.S. Highway 17 to commence firing on the splash targets, mindful that traffic was present on U.S. 17. With all the nine forward-firing guns blazing, the plane felt like it hit a brick wall, and the airspeed would drop about 50 mph. As we pulled up over the ocean in a tight right chandelle (a 180° turn with a climb), the "G" force would make it difficult to lift an arm, or if we were standing, would make us sag toward the floor of the plane. Cheeks would sag also, but as soon as we leveled and turned in our right-hand pattern over Brook Green, we would return to normal.

I keep using Brook Green as a reference point for a purpose. It contained a sort of zoo among the beautiful azaleas and tall pines, and in the midst of the garden was a round pond with a bridge across it. Always in the middle of that bridge sat a black bear. I can't count the times I lined up my gun sight on the bear with an "itchy" finger. It's the only time I have ever had the opportunity to "shoot" a bear! Occasionally, as we swung north on our flight pattern, we would have a round of ammunition get so hot in the chamber that it would "cook off," explode and send the projectile toward the city of Myrtle Beach. I don't think we ever hit anyone, but who knows?

Once on a return flight to the base, I was crawling the catwalk over the bomb bay when the plane went into violent gyrations. I was battered from top to bottom in the crawl space until we reached level flight. I proceeded to the cockpit and inquired of the pilot what problem he had encountered. He had deliberately flown through a thunderhead – just to see what it was like! That's the reason that I steadfastly claim that flying with green crews was more hazardous than combat. The pilot was a flight officer, and I chewed him out properly. Had he been a second lieutenant – I don't know.

On another occasion, we were flying to Myrtle Beach "on the deck" or "hedge-hopping" at perhaps 15 or 20 feet in altitude. It was great sport to fly up behind an automobile on one of those long straight South Carolina highways, pull up over him and drop down in front. Invariably, the driver would pull over to the side of the road – shaken. On this particular day, we found a farm family stringing tobacco in the back yard under a giant oak with the mule still hooked to the sled. As we roared over at two hundred plus mph, the mule took off, the family scattered, and dog, chickens and whatever, scattered like popcorn in a popper. All this was strictly against regulations. Our orders were to fly to Myrtle at a minimum of 1,000 feet, but the order was never enforced unless some quick thinking observer could get a number off the plane's wing. This was as much combat training as the rest. After all, the crews would make "on the deck" flights constantly overseas.

V.J. Day

Greenville was my last assignment, and it was there that I was stationed when V.J. Day was declared. From mid-August until late October, I languished at Greenville awaiting a discharge. Crews were being released and sent home, but the instructors were not. Evidently, they wanted to keep the high-point men (those with high Military Occupation Service ratings) in service just in case we were needed. Needed for what? The war was over. I had 118 points for discharge, and it only took 76 points. Eight of us who were in the same category sat and sweated it out. Radio personality H.V. Kaltenborn, bless his heart, saved the day. On his Sunday night broadcast, he asked for an explanation of why men in Greenville Army Air Base were being held when others with few or no points were being discharged. He called three or four of the names of my fellow sufferers, but not mine. The next morning, all eight of us were processed and cleared the base by nine o'clock. I never knew who had the pull, but somebody did.

The following Sunday morning at ten o'clock at Seymour Johnson Field, Goldsboro, North Carolina, I stood my last formation. I was handed my discharge with ceremony and fanfare. On leaving the base theater where the ceremony had been held, I met a captain who, upon seeing my discharge in hand and “ruptured duck” sewn on my tunic, came to sharp attention and saluted. I returned his salute and said, “At ease, Captain.”

BACK HOME

In the fall of 1945, I was separated from the Army Air Corps (it was still "Army" then) and returned to Winston-Salem to live with Mother and Dad. "Return" is hardly the word. While I had visited several times, I had never actually lived in Winston-Salem, and knew hardly anyone outside my family. I was the proverbial "lost ball in high weeds."

In order to retain a semblance of sanity, I decided to farm that fall on shares with Dad. He had a small 97-acre farm north of Winston-Salem, and it held considerable open land; ideal for raising small grain. So, I began breaking the land to sow a fall crop of wheat. Dad had an ancient F-20 Farmall tractor, but no truck. That's where I came in.

A couple of weeks or so after my discharge, and prior to beginning the wheat crop, I went to Sears-Roebuck and purchased the necessary tools and equipment consistent with farming. Following that, I read an advertisement in the local paper that veterans had first priority on the purchase of military equipment – at reduced prices. That's where I could get my farm truck!

New Wheels

Dad and I drove down to Fort Bragg, where the sale was being held, "kicked the tires" of countless vehicles until I came upon this particular 1944 Chevrolet 4-by (front wheel drive) and decided that was the one for me. It only had 3,600 miles on the odometer, and seemed in tip-top shape. The only problem: instead of a bed, it had a "fifth wheel" requiring a trailer. I didn't need a trailer; besides, it was equipped for electric trailer brakes and electric trailers were extremely rare except on military vehicles.

What the heck! I could adapt the truck and remove the "fifth wheel" and have a "stake-body" bed made for it. This I did. My cousin, Roy

Badgett, was a genius at adaptation, and I took the truck to him for fitting the bed, which he did with dispatch. Due to the short chassis because it was designed for a tractor cab, I could only fit an eleven-foot bed, but for a farm truck, it didn't matter. It would be large enough.

I brought my finished product home (it was an ugly thing, but I was proud of it) and parked it in the back yard at 2050 Elizabeth Avenue. Looking back on it, I am surprised that the neighbors didn't complain, but they didn't. I immediately put my new "contraption" into service and began hauling seed, fertilizer, and fuel to start my fall grain crop. In the meantime, my Uncle Roby Taylor came up with a brilliant idea (he had a lot of them) and commissioned me to haul a load of furniture to Rocky Mount, where his daughter (my cousin), Peggy Winstead, was setting up housekeeping in preparation for husband Ben to come home from overseas and join the comforts of civilian life. Upon delivery of the furniture, I was further commissioned to go to Petersburg, Virginia and pick up a load of cylinder oil and bring it back to Winston Salem – five tons of it! Enroute between Rocky Mount and Petersburg, I experienced my first trouble; trash in the fuel line. I had to blow out the line before proceeding to pick up the oil.

Blow Outs

What I didn't mention, and I completely ignored, was, my truck was licensed in the state of North Carolina as a farm truck and certainly not a vehicle to transport goods interstate. It did not matter to me; laws were made to be broken and somewhat foolish anyway, so I took off for Rocky Mount, delivered my load, and proceeded to Petersburg for the oil. With a full load, I started home and got about ten miles south of Petersburg when "WHAM-HISS" suddenly alerted me of big trouble – blowout! Fortunately, it was one of the dual rear wheels, and I had a spare. I do not recall who or how the tires got changed, but they did. In the meantime, I discovered that all six wheels, plus the spare, were equipped with old retread tires – dry rotted to boot!

I could not do anything but proceed toward Winston-Salem and hope for the best. All was going well until just out of Clarksville, Virginia, another HISS and a flat tire. My load was far in excess of the ability of my tires to carry it. I limped (tip-toed was more like it) into Clarksville to have my second change of tires. By this time, I knew I was in serious trouble. As the station attendant jacked the rear of the vehicle sufficient to change tires, his jack started sinking through the asphalt apron. "What do you have on this truck," he asked, "lead?" At any rate, he did get the tire changed, and I paid his fee, leaving me with little cash for the balance of my trip.

I got back in the truck and started for Danville, Virginia, after having a lunch of a pack of nabs and a Coke. As I approached Danville, you guessed it, another WHAM, and a blown tire. I mean this one was blown but good – a hole in the sidewall you could stick your fist through! Until that moment, I had never even heard of Murphy's Law, much less believed in it. I got educated in a hurry. I got a garage to come fix the tire with a "boot" and exchanged it with another tire that hadn't blown and put the "booted" tire on the front wheel, where the weight would be considerably less. As I left Danville, I had only a Coke and nabs since breakfast, and thirty-seven cents in my pocket. Fortunately, I did have enough money to buy enough gasoline to get me home.

At this point, I had changed plans and routed my travels by the most direct route to my home, which entailed roads having gross-load restrictions much less than the tonnage I was carrying. Here I was: hungry, no money, three bad tires and who knows how many more would blow before reaching my destination. At that point, I asked the Lord if he would let me get this "albatross" home, I would never make another trip in it. He heard me, for as I approached the Virginia/North Carolina state line, I passed a State Trooper who paid no attention to me. The rest was clear sailing!

Upon arriving home in the wee hours of the morning, I parked on the street in front of the house and went inside. Mother and Dad were awake, probably wondering what had become of me. He asked, "Don't you think you should go down to Roby's plant and unload? All that weight will be hard on the springs." To which I replied, "I do not care what happens to that truck," and went to bed.

I did use it as a farm truck for the balance of the fall, and it came in quite handy as transportation that winter since it did have four-wheel drive. It had served its original purpose and was faithful to a fault!

A Good Investment

I had purchased the vehicle for something in the neighborhood of $750, and sold it in the following spring for approximately $1,100, on top of which I was paid $150 for the Petersburg trip – a pretty good return on investment for those days. I sold the truck to a man who was in the business of hauling stove wood in the black section of Winston-Salem, which he put to good use, I'm sure. My point is: I followed my commitment to the Lord. I went out of the trucking business!

MILITARY RECORD
GLENN E. TAYLOR ASN 34 770 263

Inducted	Camp Croft, SC	03/20/43
PFC	Keesler Field, MS	04/28/43
Sgt	Harlingen AFB, TX	11/27/43
S/Sgt	Attlebridge, England	06/07/44
Disch	Seymour Johnson, NC	10/22/45

MISC OTHER BASES:
British Isles:
Stone, England
Clonto & Greencastle
Bungay (446th BG)
Attleboro (Rep. Depot)

ROUTE OVERSEAS

Hamilton Field, CA	4/02/44
Sky Harbor, Phoenix, AZ	4/02/44
Midland, TX	4/04/44
Warner Robbins, GA	4/05/44
Morrison Field, FL	4/06/44
Morrison Field, FL (Departure)	4/12/44
Waller Field, Trinidad	4/13/44
Belem, Brazil	4/14/44
Fortaleza, Brazil	4/15/44
Fortaleza, Brazil (Departure)	4/17/44
Dakar, French West Africa	4/18/44
Marrakesh, Morocco	4/19/44
Marrakesh, Morocco (Departure)	4/21/44
Valley, Wales	4/21/44

USA
Fort Bragg, NC
St. Petersburg, FL
Hammer Field, CA
Muroc (Edwards AFB)
Miami Beach (R & R)
Laredo, TX
Charleston, SC
Greenville AAF, SC
Eglin AFB, FL
Myrtle Beach, SC

MISSIONS RECORD, 466TH BOMB GROUP, ATTLEBRIDGE, ENG; STA.120

784th Squadron				785th Squadron	
1	Melun, FR	05/24/44	17	Eisenach, GR	07/20/44
2	Mulhouse, FR	05/25/44	18	Munich, GR	07/21/44
3	Tutow, GR	05/29/44	19	St. Lo, FR	07/25/44
4	Pas de Calais, FR	06/02/44	20	Bremen, GR	07/29/44
5	Abbeville, FR	06/03/44	21	Remaisnil-Bellow Sur Somme, FR	08/02/44
6	Port en Bessin, FR	06/06/44	22	Le Grandee Valley	08/03/44
7	Port en Bessin, FR	06/07/44	23	Rostock, GR	08/04/44
8	Chateaudum, FR	06/10/44	24	Brunswick, GR	08/05/44
9	Illers, FR	06/12/44	25	Claistres, FR	08/08/44
10	Pas de Calais, FR	06/14/44	26	Saarbrucken, GR	08/09/44
11	Westermunde, GR	06/18/44	27	Strasbourg, GR/FR	08/11/44
12	Pas de Calais, FR	06/19/44	28	Mourleon, FR	08/12/44
13	Pas de Calais, FR	06/23/44	29	Hanover, GR	08/24/44
14	Cauchie D'Eiques, FR	06/24/44	30	Oraineburg, GR	08/27/44
15	Bretagne, FR	06/25/44	31	Pas de Calais, FR	08/30/44
16	Troarn, FR	07/18/44	32	Karlsruhe, GR	09/05/44

Boarded Queen Mary 10/23/44 in Glasgow, Scotland, for return to USA. Arrived in New York 10/28/44

SECTION III: CURRENT DAY WWII MEMORIES

Sixty years ago, our nation was awakened to the reality of war. On December 7, 1941, the Japanese perpetrated an unprovoked attack on Pearl Harbor, thus forcing our peace-loving nation to recognize the error of its idea that "whatever happened in Europe was not our problem." We were involved at a moment's notice and were unprepared for it!

As a student at North Carolina State College, studying pre-aeronautical engineering, I was just completing my first semester and "cramming" for final exams, when one of my classmates came into the library and loudly proclaimed, "The Japs have just bombed Pearl Harbor." I had never even heard of Pearl Harbor and knew nothing of its location or the significance of the action. Studies forgotten, we returned to the dormitory to lose ourselves in the latest war news. We all knew that this meant we would be involved, but had no idea of which direction fate would lead us.

After more than a year of waiting for a call, and still a student at State, I was finally summoned to report for induction at Camp Croft, Spartanburg, South Carolina. Having been accepted in the Army Air Corps, I followed the usual routine of basic training, engineering school, gunnery school and phase training prior to departing for my overseas combat assignment, all of which took a little more than a year.

My crew and I left Morrison Field, West Palm Beach, Florida, on Easter Monday morning in our new B-24 bomber. Our destination was with the Eighth Air Force in England, via Trinidad, Brazil, Morocco, French West Africa, and Valley, Wales. We arrived in late April of 1944 and, following further schooling in escape and evasion, we finally settled in our base at Attlebridge (station 120) near Norwich. My first mission was May 24th and the last was September 5th. In that period, I flew 32 missions, mostly in France and Germany.

At the time, we were told that we might as well consider ourselves dead, as the statistics were weighted against our survival. A three percent loss of airplanes on each mission was considered "acceptable," and we flew 32 missions. Multiplying that out meant that we had an almost 100% chance of being shot down! And that wasn't the worst of it; we found out in later years.

World War II statistics indicate that there were 210,000 airmen who flew with the Eighth during the three years we were engaged in Europe. Statistics also show that 26,000 were killed in combat (12%) and 28,000 prisoners of war or escapees (13.3%). When grouping these two together, we suffered and astounding 25+% casualty rate. One fourth of our flyers did not return to their English base.

I am now (in 2001) seventy-seven years of age, of which almost 57 years have passed since I completed my tour of duty. I consider that 57 years of grace.

You think I'm not glad to be here!

THE ROCKET BOYS

While living in Dothan, Alabama in the late 1950s and early '60s, my wife, Sarah Jane, and I were invited to attend the Panama City, Florida annual Chamber of Commerce banquet featuring the famous WWII German rocket scientist Dr. Werner Von Braun. Dr. Von Braun's message that evening was how the Redstone Arsenal team was planning a trip to the moon on a Saturn rocket. What a fascinating concept that became a reality a few years later.

At that meeting, I met a young member of the Von Braun team by the name of Walter Weisman, who also came from Germany following World War II to help develop the rocket program. Walt was not a scientist, but with his warm personality and out-going mannerism, was a natural public relations spokesman. We exchanged pleasantries that evening, but I do not remember any particular bonding other than he was a member of "THE ROCKET TEAM."

Several years passed, and my family and I moved to Birmingham where I became the Director of the Committee of 100 and later Resident Vice President of Southern Railway. In that position, I was invited to join the Birmingham Kiwanis Club and placed on its Program Committee. I volunteered to exercise my contact with Weisman to invite him to speak to our Club on the subject of the missile program. This was near the time our space program was about to land on the moon, as predicted by Dr. Von Braun several years earlier.

A Small World

Walt accepted my invitation and delivered a meaningful and detailed program describing the rocket and moon landing. His message was well received, and I invited him to The Club for refreshments after the Kiwanis meeting. We had a two or three-hour wait before his flight returned him to Huntsville, so we visited at some length at the bar and exchanged stories of our youth and war.

Sergeant Weisman served with the Luftwaffe during WWII and was a member of Von Braun's staff at Peenemunde, Germany while I, of course, was a Staff Sergeant in the U.S. Eighth Air Force. We were bound to have some interesting coincidences. We did! On my third mission to Tutow, Germany, which is on the northern outskirts of Berlin, I flew as ball gunner. Upon completing the bomb run, we relaxed a bit and flew on a northerly course directly over Peenemunde, not knowing that this was the site where the V-I and V-II rockets were being developed. To say we received a rough welcome would be a gross understatement! Over the Tutow target, we had received only light and inaccurate flak. However, when we reached the skies above the rocket development laboratory, all hell broke loose! Flak was intense and extremely accurate. We picked up a few holes in our ship, and I recovered a piece of flak about three inches long and extremely jagged, which I still have.

As Walt and I compared experiences, we began to develop a camaraderie that lasted as long as he lived. I have fond memories of our many subsequent visits. As we concluded our visit at The Club, laced with the warmth of a couple of shots of bourbon, we commented on how strange this world really is. Here sat two old has-been warriors who fought on opposite sides of the worst single conflict in the annals of world history. Suddenly, we found that we were not only good friends, but had so much in common. Thirty-some years earlier, we would have tried to kill each other without giving it a second thought. As a matter of fact, we HAD tried to do just that!

Some years later, we attended a meeting in Montgomery where the scheduled speaker's plane was delayed by fog. After dinner while awaiting the speaker's arrival, the Master of Ceremonies asked Walt to entertain the gathering as a time-filler. In his impromptu remarks, Walt made several references to our friendship. Later during a break, someone asked me how Walt and I became such good friends. I replied, "We flew with the Luftwaffe together!"

Alabama Memorial Committee at the Gathering of Eagles at the Mighty Eighth Air Force Heritage Museum near Savannah, Georgia. Committee members L to R: unknown, unknown, Major General Lew Lyles, Glenn Taylor, John Nixon, Frank Brewer, C.B. "Red" Harper, Henry Arnold, Lt General Buck Shuler, Charlie Beard.

Alabama Chapter Memorial marker designed by Author, located at the Mighty Eighth Air Force Heritage Museum, Savannah, Georgia

Veterans from the Alabama Chapter of the Eighth Air Force Historical Society attending a static display of WWII aircraft. L to R: Benton White, Glenn Taylor, Henry Arnold

B-24 on static display at the Shelby County (AL) Airport. Display sponsored by the Collins Foundation

Mighty Eighth Air Force Heritage Museum, Savannah, Georgia

Chapel of the Fallen Eagles at the Heritage Museum

Alabama Chapter of the 8th Air Force Historical Society Memorial Marker designed by Author. Memorial is located at the American Village, Montevallo, Alabama

PATRIOTISM

Recently while my wife and I were driving home from church, I noticed a late model SUV proudly flying the American flag from a staff over the right front door. My thoughts reflected back to the time after 9/11 when so many automobiles displayed this emblem of patriotism and how seldom it is seen presently. My thoughts were interrupted as the small flag, waving so proudly at fifty miles an hour, suddenly flew off the car. I didn't see it land, but it struck me as ironic that we so often find our patriotism landing in the side ditch when we feel safe, when the urgency of danger has passed.

To me, patriotism is much more than flag waving, although I still get choked up when I see our beautiful flag pass in parade while our national anthem is being played. I notice, all too often, that people stand slouched and unconcerned without the right hand over the heart when the flag passes. That bothers me.

I have visited American veterans' cemeteries throughout our country and a couple in Europe. All those headstones present a vivid picture of the great sacrifices our men and women have endured through our county's short history to give us the freedoms we enjoy today: freedom from fear, from want, from oppression, as well as the freedom to NOT salute the flag when it passes.

On both the 40th and 50th anniversaries of the Normandy Invasion, I stood on the beaches of Normandy, France. Later I walked through the American cemetery at St. Laurent among the 9,600 crosses and Stars of David, experiencing the sadness of the loss of those lives sacrificed on June 6, 1944. Having been a participant in the air campaign that helped protect the brave men that hit the beaches that morning, I will carry the awesome spectacle of that invasion armada with me as long as I live.

In our country's relatively brief history, over two million men and women have died in the defense of our freedoms. What a heinous price to pay! Yet, we could not be the nation we are today without the blood sacrifice of those who have gone before us. They are the patriots we should salute today and every day.

Since 9/11, I have proudly flown our national flag in my front yard – day and night – in support of our troops abroad, our President and his cabinet members who are prosecuting the war against terrorism with the ultimate goal of victory. I don't apologize for being a patriot. I don't think it is sissy to choke up and shed a tear when filled with emotion over national pride. And, yes, I do still get choked up each time I hear Kate Smith's rendition of "God Bless America."

EPILOGUE

By the Fall of 1945 I was "OUT of the service!" Out for what? For the most part, I found my time in the service to be both rewarding and exciting, but now I was on my own. No one to account to; no one to account for.

I spent my early months after release from the Army trying to sort out my future and decide what direction to follow. The GI Bill was available, but a lot of job opportunities existed for returning veterans. After working in a cotton mill for a few months as a Time/Motion Apprentice, I decided, on good advice, to enroll in college. After three years at the University of North Carolina, I received my diploma and accepted a job with Ford Motor Company. The following year, I moved to Fayetteville, N.C., as assistant manager for the local Chamber of Commerce.

In 1952, I was offered a job with the Albemarle, N.C. Chamber Commerce. I served there for five years as Manager and then moved to Dothan, Alabama, for six years as Executive Vice President. While I enjoyed those assignments, I wasn't achieving the financial security for which I had hoped. Neither job held any promise for the future, such as medical insurance or retirement benefits. I knew that sooner or later I would need those – plus sufficient funds to send my three children through college.

Then, opportunity arrived. In 1963 my family and I had moved to Birmingham, Alabama, as Director of the Committee of One Hundred (the industry-hunting organization of greater Birmingham), and I began my quest for new and expanding job opportunities. But, the roof caved in! The Civil Rights "Freedom Marches" in the 1960's all but shut Birmingham down, and the chances of attracting new businesses and industries to the area were virtually non-existent.

In 1968, I was contacted by a representative of Southern Railway, who was exploring the possibility of me joining their Industrial Development Department. Little did I know that railroads take their time deciding on anything, so after waiting for several months and not hearing anything more about the job, I dismissed the idea.

"Out of the blue" one day, I was asked to have lunch with Southern Railway's President, D.W. Brosnan. At that time, he offered me the job as Resident Vice President for Alabama – and I took it!

Upon my retirement after twenty-three years with the railroad (Southern and Norfolk-Southern), I must say that that position fulfilled all the aspirations I had ever sought. Moreover, I found "The Railroad" to be the answer to my life's goals. Not perfect, but extremely rewarding.

Glenn E. Taylor
2011

ABOUT GLENN E. TAYLOR

In 1931, when he was seven, Glenn Taylor's family moved to Fairntosh Plantation near Durham, NC. Leaving the farm for college in 1941, Glenn studied aeronautical engineering until joining the Army Air Corps in 1943. Stationed in England with the 8th Air Force, he flew 32 missions, May 24-September 5, 1944. After discharge in 1945, Glenn earned a Journalism degree from University of North Carolina in 1950. For 16 years, Glenn managed Chambers of Commerce in the Southeast and spent 23 years with Norfolk Southern Railway, retiring in 1987. Currently living in Birmingham, Alabama, Glenn is married to Sarah Jane, has three children and six grandchildren.

CPSIA information can be obtained at www.ICGtesting.com
Printed in the USA
LVOW081339301111
257159LV00005B/1/P

9 780983 987666